THE AMAZING BRAIN

The Amazing Brain

Robert Ornstein and Richard F. Thompson
Illustrated by David Macaulay

HOUGHTON MIFFLIN COMPANY/BOSTON

Library of Congress Cataloging in Publication Data

Ornstein, Robert Evans

The amazing brain.

1. Brain I. Thompson, Richard Frederick, date.

II. Title.

QP376.076 1984 612'.82 84-12907

ISBN 0-395-35486-2

ISBN 0-395-58572-4 (pbk.)

Printed in the United States of America

CRS 14 13 12 11 10 9 8 7 6

Houghton Mifflin Company paperback 1986

Contents

Preface

FOR THOUSANDS OF YEARS people have tried to understand the brain. The ancient Greeks thought it was like a radiator, to cool the blood. In this century it has been compared to a switchboard, a computer, and a hologram, and no doubt it will be likened to any number of machines yet to be invented. But none of these analogies is adequate, for the brain is unique in the universe, and unlike anything man has ever made.

Over the past few decades, great advances have taken place in the various fields of study that touch upon the brain. From evolutionary biology we have learned how and when the different parts of the brain were "built." From neuroanatomy we know how the elements of the brain are assembled, and from neurophysiology we have begun to understand how those elements and the chemicals that make them up function together. We are now beginning to understand what or "who" the brain is, but a great deal remains to be discovered in the neurosciences.

The brain is like an old ramshackle house that has been added on to over the years in a rather disorganized fashion. In this book, we look at the architecture of that house, first by taking a tour through the various "rooms" and then by going deeper and deeper into the material with which those rooms are constructed. Later we consider some of the mysteries of the brain and human experience. The drawings and diagrams appearing throughout will help you to visualize some of the more complex aspects of the amazing organ that is the human brain.

The Amazing Garden

or

A romantic view of the evolutionary growth of the brain with appropriate anatomical diagrams

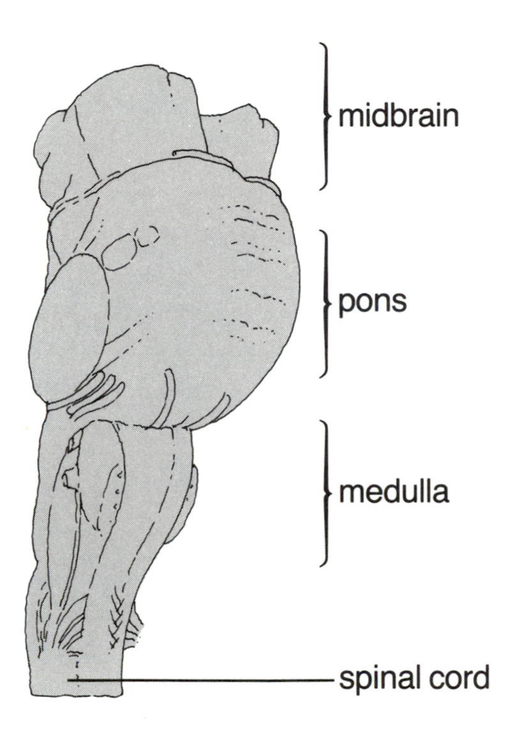

The Brainstem

The brainstem is the oldest part of the brain. It evolved more than five hundred million years ago. Because it resembles the entire brain of a reptile, it is often referred to as the reptilian brain. It determines the general level of alertness and warns the organism of important incoming information, as well as handling basic bodily functions necessary for survival—breathing and heart rate.

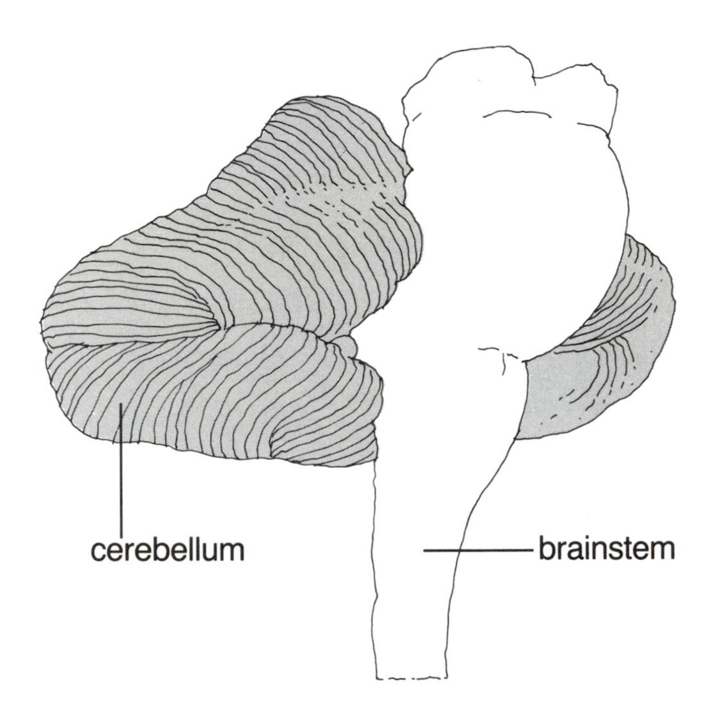

The Cerebellum

The cerebellum, or "little brain," is attached to the rear of the brainstem. Among other functions, the cerebellum maintains and adjusts posture and coordinates muscular movement. The importance of these functions is evident when we realize that the cerebellum in the human brain has more than tripled in size in just the last million years. It now appears that memories for simple learned responses may be stored there.

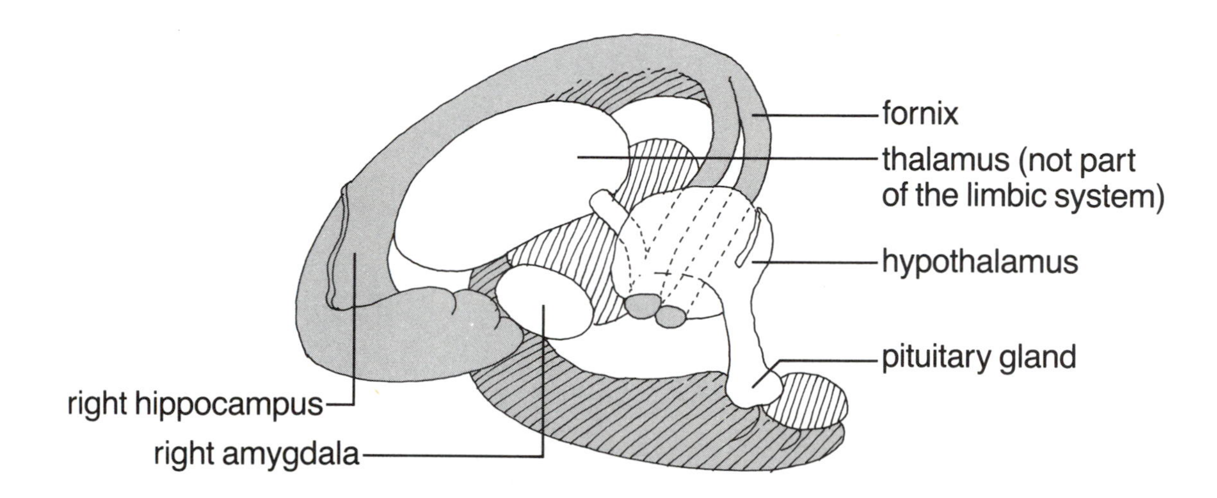

The Limbic System

The limbic system is the group of cellular structures located between the brainstem and the cortex. It evolved sometime between three hundred and two hundred million years ago. Because the limbic system is most highly developed in mammals, it is often called the mammalian brain. In addition to helping maintain body temperature, blood pressure, heartbeat rate, and blood sugar levels, the limbic system also is strongly involved in the emotional reactions that have to do with survival.

Two key parts of the system are the hypothalamus (below the thalamus) and the pituitary gland. Although only about the size of a pea, the hypothalamus regulates eating, drinking, sleeping, waking, body temperature, balance, and many other functions. Through a combination of electrical and chemical messages, it directs the pituitary gland—the master gland of the body.

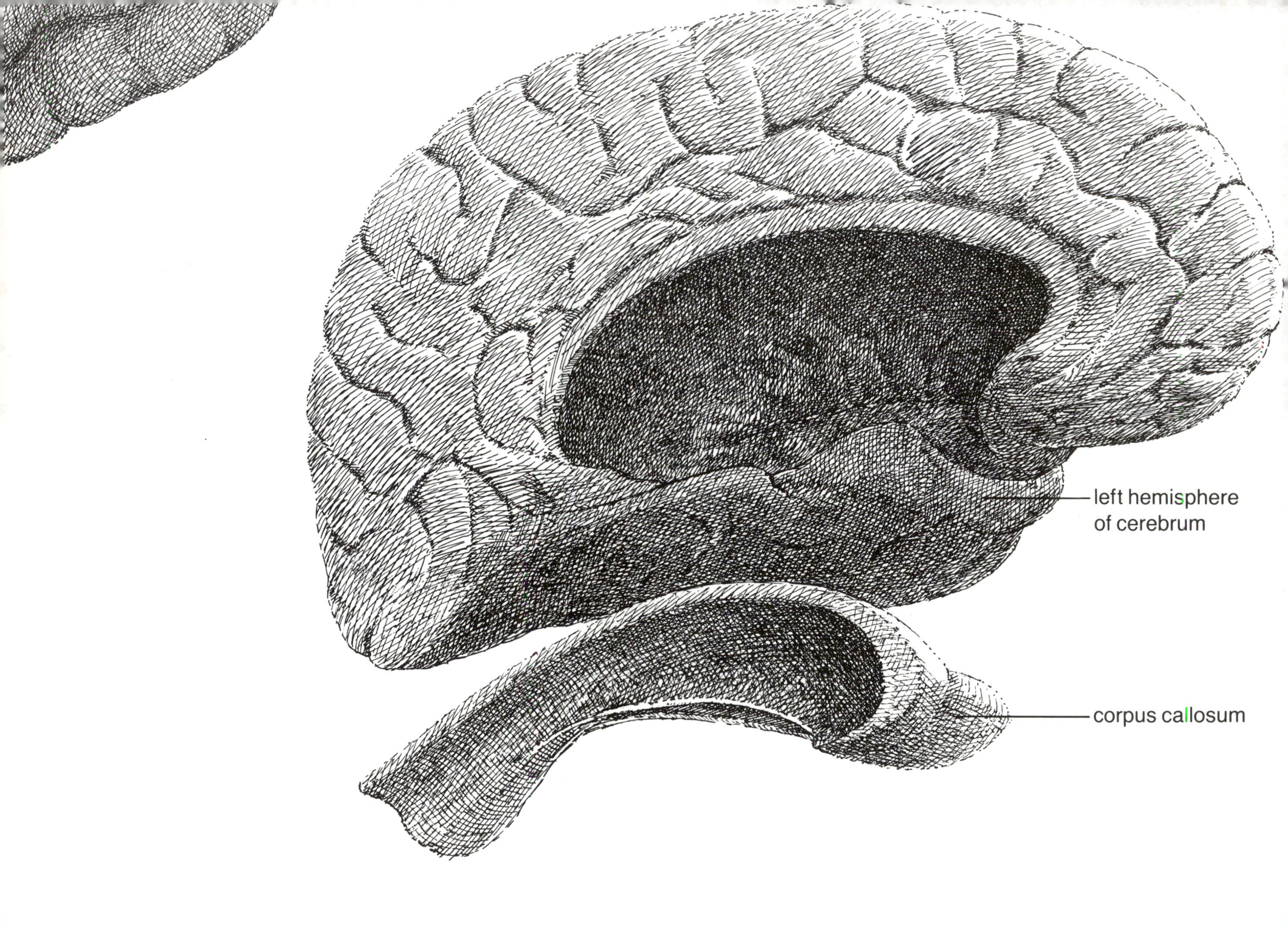
left hemisphere
of cerebrum
corpus callosum

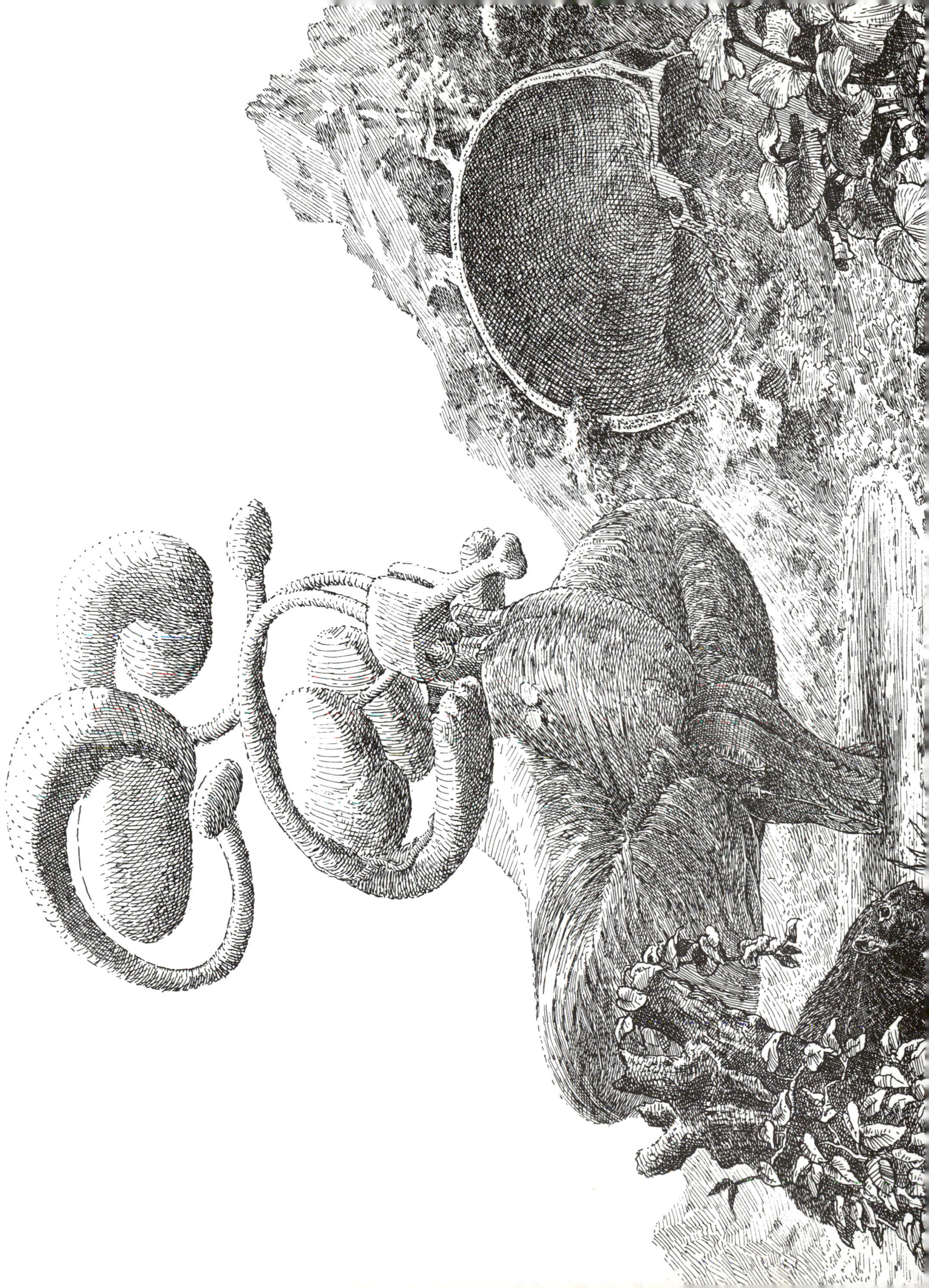

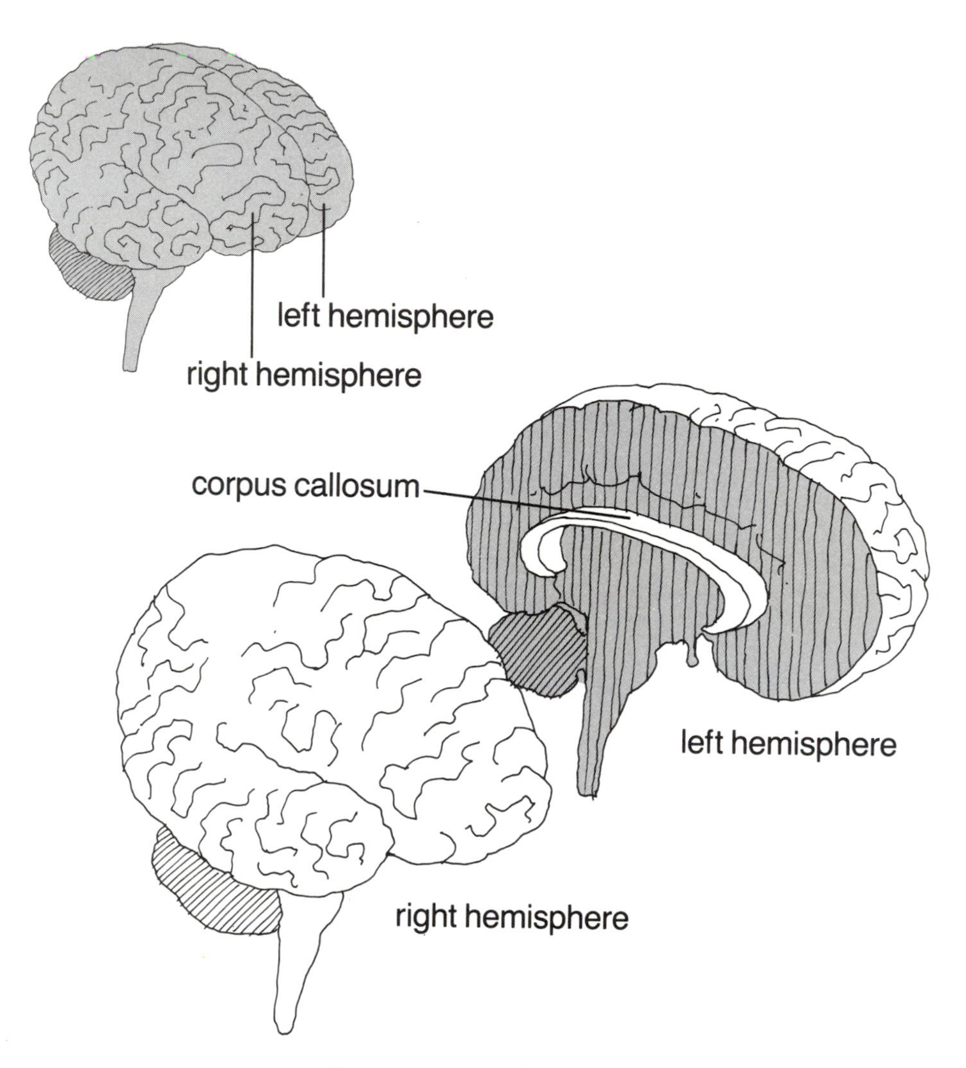

The Cerebrum

The largest part of the human brain is the cerebrum. It is divided into two halves, or hemispheres, each of which controls its opposite half of the body. The hemispheres are connected by a band of some three hundred million nerve cell fibers called the corpus callosum. Covering each hemisphere is a one-eighth-inch-thick, intricately folded layer of nerve cells called the cortex. The cortex first appeared in our ancestors about two hundred million years ago, and it is what makes us uniquely human. Because of it, we are able to organize, remember, communicate, understand, appreciate, and create.

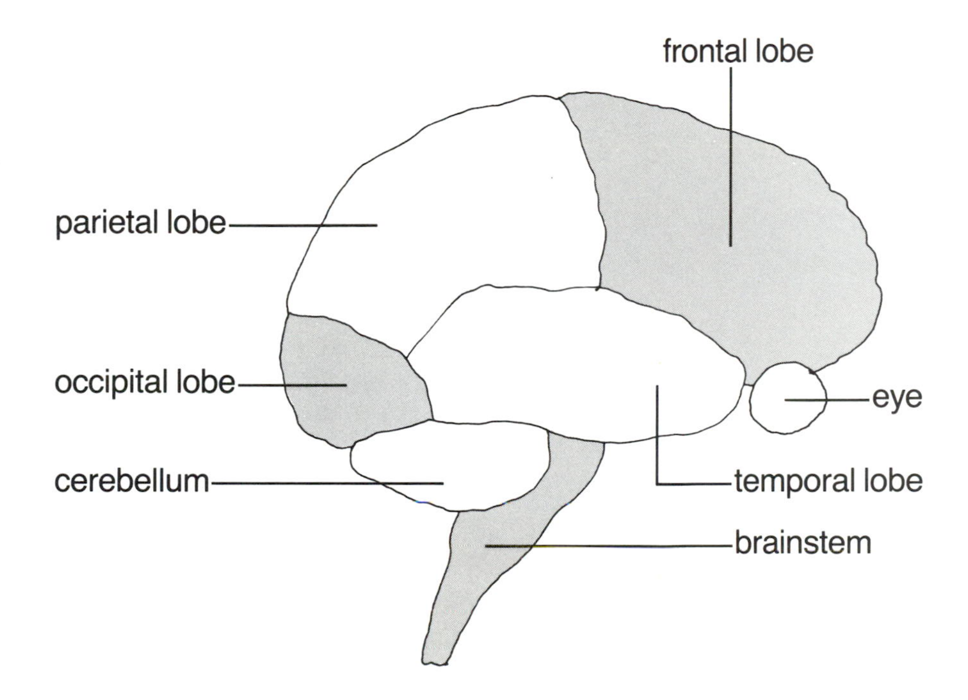

The Lobes of the Cortex

The cortex of each hemisphere is divided into four areas called lobes. The frontal lobe is primarily involved in planning, decision making, and purposeful behavior. The parietal lobe represents the body in the brain. It receives sensory information from the body. Part of the occipital lobe is devoted to vision and is often called the visual cortex. The temporal lobe appears to have several important functions, including hearing, perception, and memory.

WANDELAAR
ET NETTER

PART I

THE RAMSHACKLE BRAIN: THE ROOMS, COLUMNS, BRICKS, AND CHEMICALS

1

The Architecture of the Brain

IT IS ABOUT the size of a grapefruit.

It weighs about as much as a head of cabbage.

It is the one organ we cannot transplant and be ourselves.

The brain regulates all bodily functions; it controls our most primitive behavior — eating, sleeping, keeping warm; it is responsible for our most sophisticated activities — the creation of civilization, of music, art, science, and language. Our hopes, thoughts, emotions, and personality are all lodged — somewhere — inside there. After thousands of scientists have studied it for centuries, the only word to describe it remains *amazing*.

There are perhaps about one hundred billion neurons, or nerve cells, in the brain, and in a single human brain the number of possible interconnections between these cells *is greater than the number of atoms in the universe.*

Although we may never completely unravel the mysteries of the brain, we do know a lot about it. We know something of what the brain is, what it does, and how it got that way.

Here is a way to help you visualize it: Place your fingers on both sides of your head beneath the ear lobes. In the center of the space between your hands is the oldest part of the brain, the brainstem. Now, form your hands into fists. Each is about the size of one of the brain's hemispheres, and when both fists are joined at the heel of the hand they describe not only the approximate size and shape of the entire brain but also its symmetrical structure. Next, put on a pair of thick gloves — preferably light gray. These represent the cortex (Latin for "bark") — the newest part of the brain and the area whose functioning results in the most characteristically human creations, such as language and art.

There is a kind of architecture of the brain, because the brain was constructed in a particular manner, by the processes of evolution, over millions of years. And so we think of it as like a ramshackle house that was built long ago for a small family and then added on to, over several generations of growth. The original structure remains basically

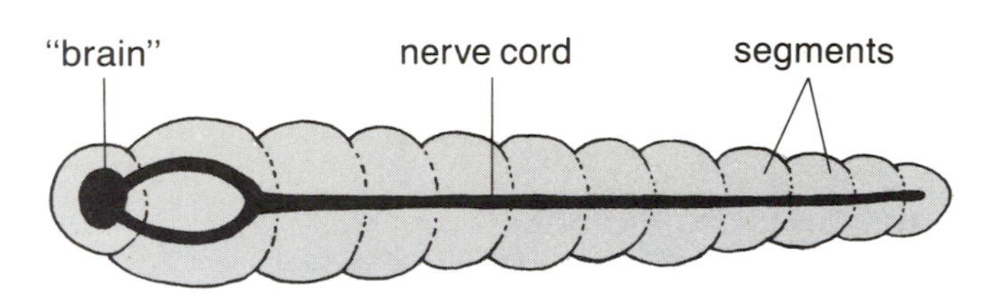

The Nervous System of the Worm

The basic plan of the human nervous system can be seen in even the most primitive animals that are bilaterally symmetrical, such as worms. These animals all have a head end and a tail end and a left and a right side.

The bodies of all such animals are segmented. Each segment contains bundles of nerve fibers that send information from receptor cells in the skin to a group of nerve cells that in turn projects information back to control the muscles. The various groups of nerve cells, called nuclei or ganglia, communicate with one another through larger bundles of nerve fibers that extend up and down the body, forming the nerve cord. When the first vertebrates developed from invertebrate ancestors, the nerve cord became encased in a bony covering, the spinal vertebrae, and became the spinal cord.

Though the human body does not seem to have segments, the human spinal cord does. It has many segments, each receiving information from a particular region of skin and controlling the underlying muscles. The nervous system of the most primitive vertebrates was little more than a spinal cord with a small enlargement at the head end. What exists as only a few extra cells in the head of the earthworm, handling information about taste and light, has evolved in us humans into the incredibly complex and sophisticated structure of the human brain.

intact, but some of the original functions have moved elsewhere in the house, as when a new, modern kitchen is built, and the old one remains but is now a pantry. So it is with the lower—that is, original—structures of the human brain, the old rooms. The upper layers of the brain, those that came later, are also very different. The brain is not like a sleek, modern house, one with each cubic foot well organized. It is chaotic, it consists of "room" upon "room" of different structures that communicate with each other through special pathways designed to keep them in touch.

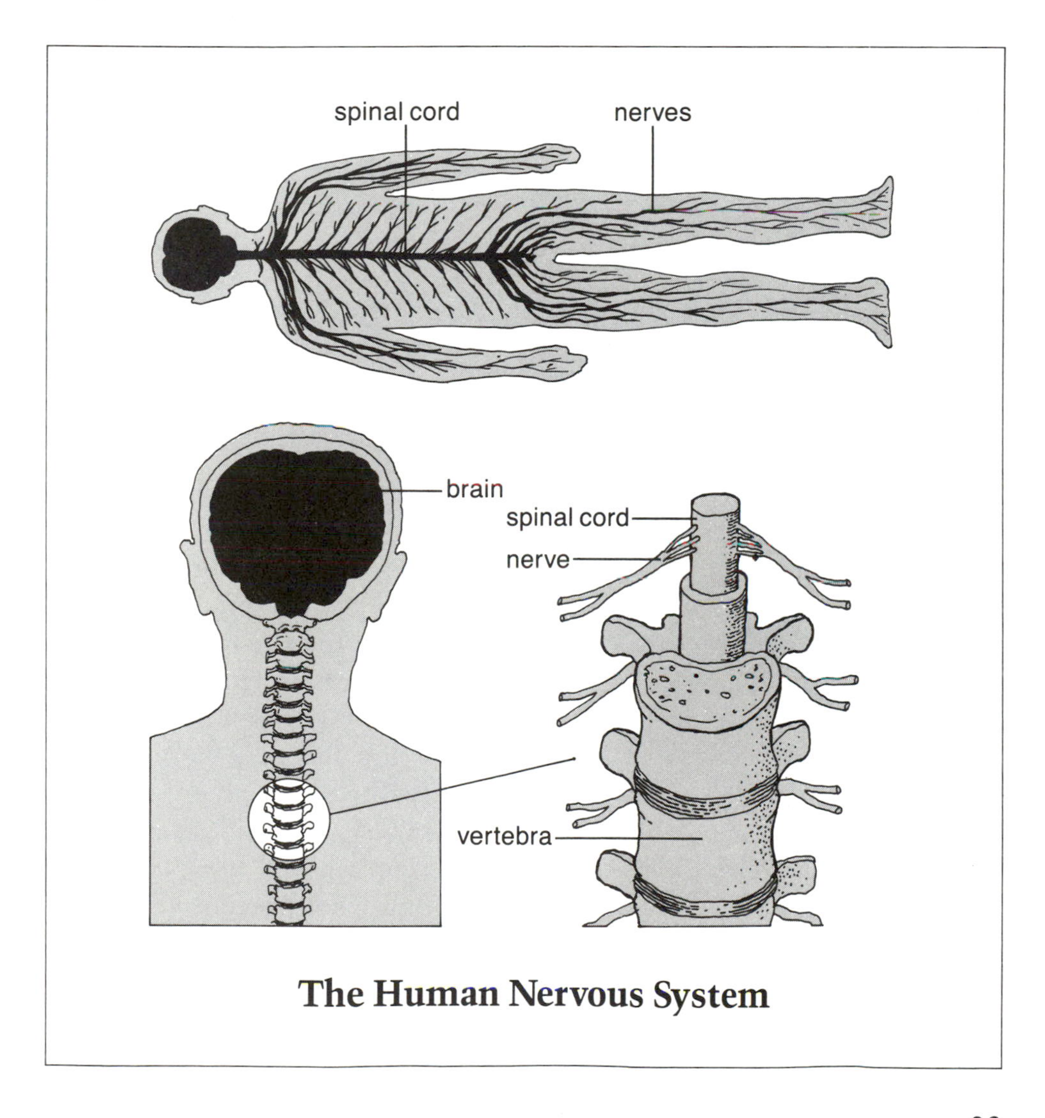

The Human Nervous System

We often like to think we live exclusively in the modern, rationally designed rooms of the house, but this is an illusion. Several systems within the brain are watching different parts of our world, coming to different conclusions, and acting in ways reminiscent of our past. Our hair may stand on end when we are angry at a business associate, a vestige of the way the fur of a cat fluffs up to increase its apparent size to an enemy.

We carry our evolution inside us, in the different structures of our brain, built in different eras.

Emotions were here before we were.

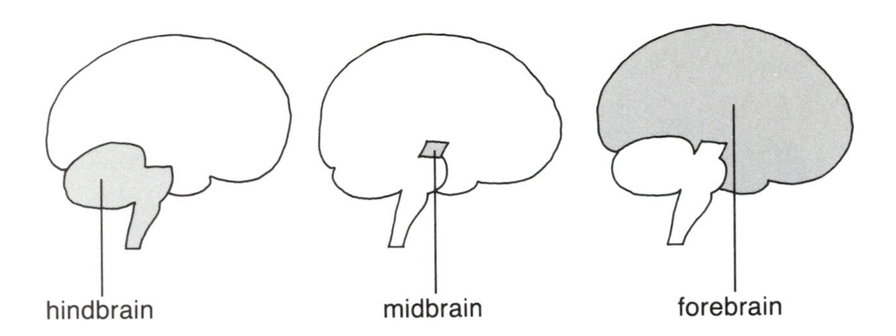

The brain was "built" in three somewhat generalized divisions. The first is the hindbrain—the oldest part of the brain—and it includes most of the brainstem. The second is the midbrain, which is only the uppermost portion of the brainstem. The third is the forebrain, which, while containing some older structures, is primarily made up of the most recently evolved areas of the brain, including the cortex.

Now, let's take a tour of our ramshackle house, looking first at those "rooms" that developed to keep us and our ancestors alive, and then at those concerned with creating new life and new worlds.

We begin with the brainstem—basically like the small enlargement at the front end of the spinal cord in primitive vertebrates. It is the oldest and deepest area of the brain, having evolved more than five hundred million years ago—before mammals. Many scientists refer to the human brainstem as the reptilian brain, because it looks somewhat like the entire brain of a reptile. The brainstem is primarily concerned with life support—the control of breathing and heart rate.

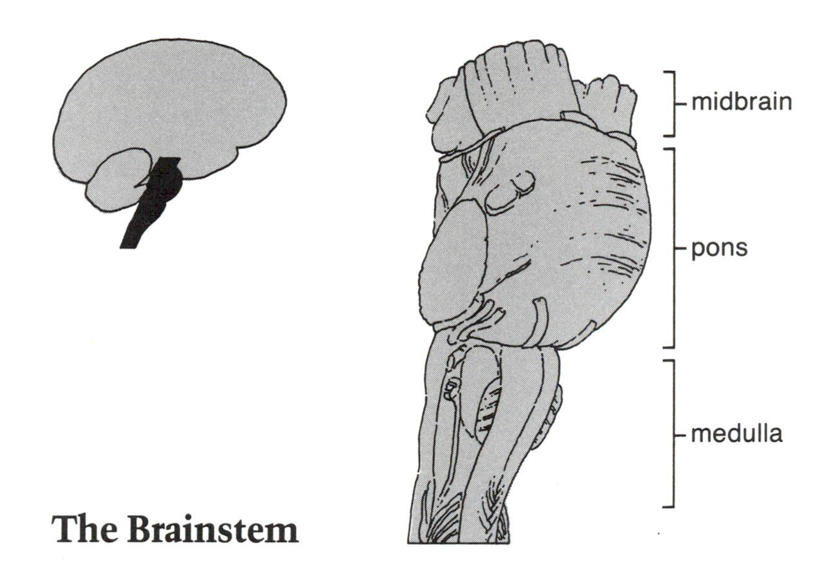

The Brainstem

Located in the center of the brainstem and traveling its full length is a core of neural tissue known as the reticular formation. It contains a number of nuclei, which are part of the reticular activating system, or RAS. Like a telephone bell, the RAS alerts the cortex (the thinking area) in a general way about arriving information (such as "visual stimulus on its way").

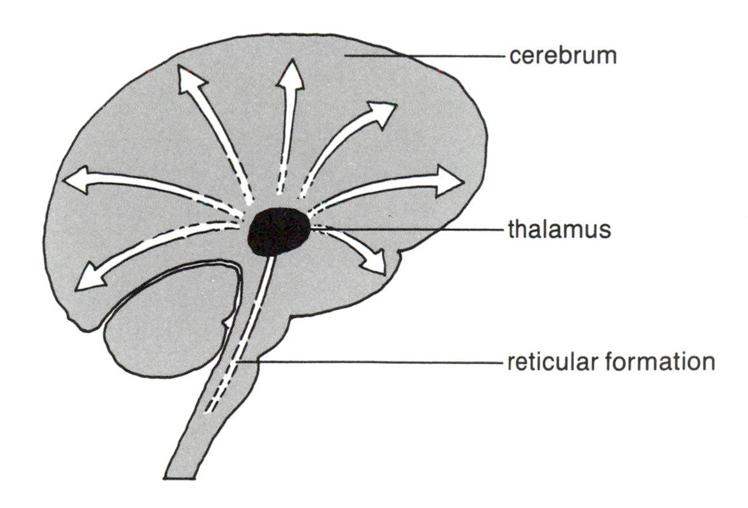

The Reticular Activating System (RAS)

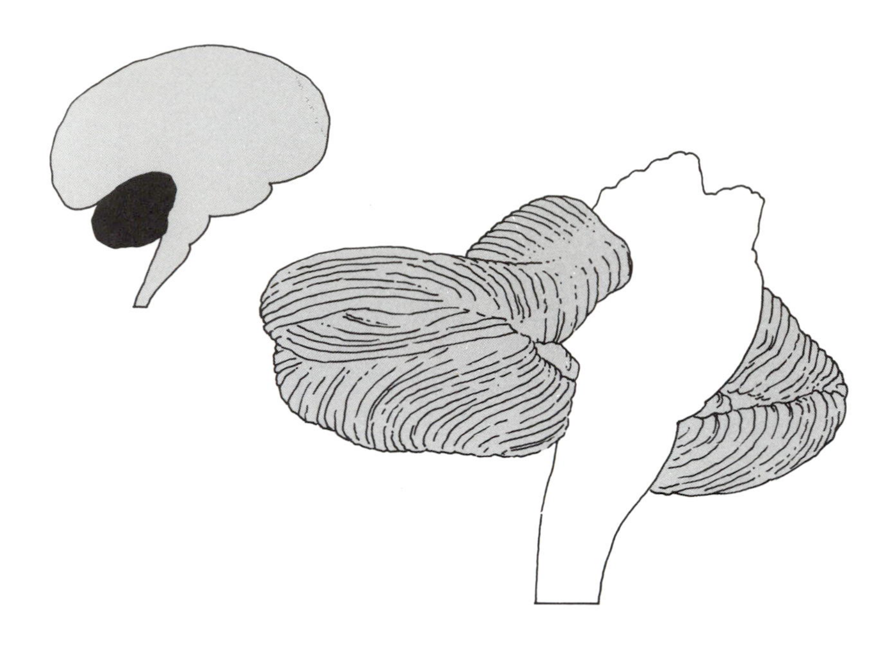

The Cerebellum

Attached to the brainstem is another part of the hindbrain, called the cerebellum. It initially developed as a "motor" structure to improve control of balance, body position, and movement in space. It now appears that memory for certain types of simple learned responses may be stored there, particularly in those parts of the cerebellum that evolved most recently. The assignment of new responsibilities to the cerebellum is typical of the way the brain has evolved. Old structures were not discarded but rather expanded to handle other functions. As more neural tissue was added to the cerebellum, a portion of the brainstem called the pons ("bridge") developed just below the midbrain to relay information to and from the cerebellum.

The limbic system is a group of cell structures in the center of the brain, immediately above the brainstem. It evolved somewhere between three hundred million and two hundred million years ago. Much of the reptilian forebrain is limbic system, dominated by olfactory input (sense of smell). In the reptile it serves as the "highest" region of the brain. In the human brain, the limbic system is dwarfed by more recent structures, and it seems to have assumed rather different functions. Olfactory input is not so important for us, and the limbic system has come to play a key role in storing memories of our life experiences, making it another example of evolution "remodeling old rooms to serve new functions."

The limbic system is often called the mammalian brain because it is most highly developed in mammals. It is the area of the brain that helps to maintain homeostasis, or a constant environment in the body. Homeostatic mechanisms located in the limbic system regulate such functions as the maintenance of body temperature, blood pressure, heart rate, and levels of sugar in the blood. Without a limbic system we would be like "cold-blooded" reptiles: we couldn't adjust our internal state to maintain its constant "climate" in spite of differing external conditions of heat and cold. Though a person in a coma has temporarily lost the use of those portions of the forebrain required for responding to or interacting with the outside world, he or she still continues to live because the brainstem and the limbic system maintain and regulate vital body functions. The limbic system is also strongly involved

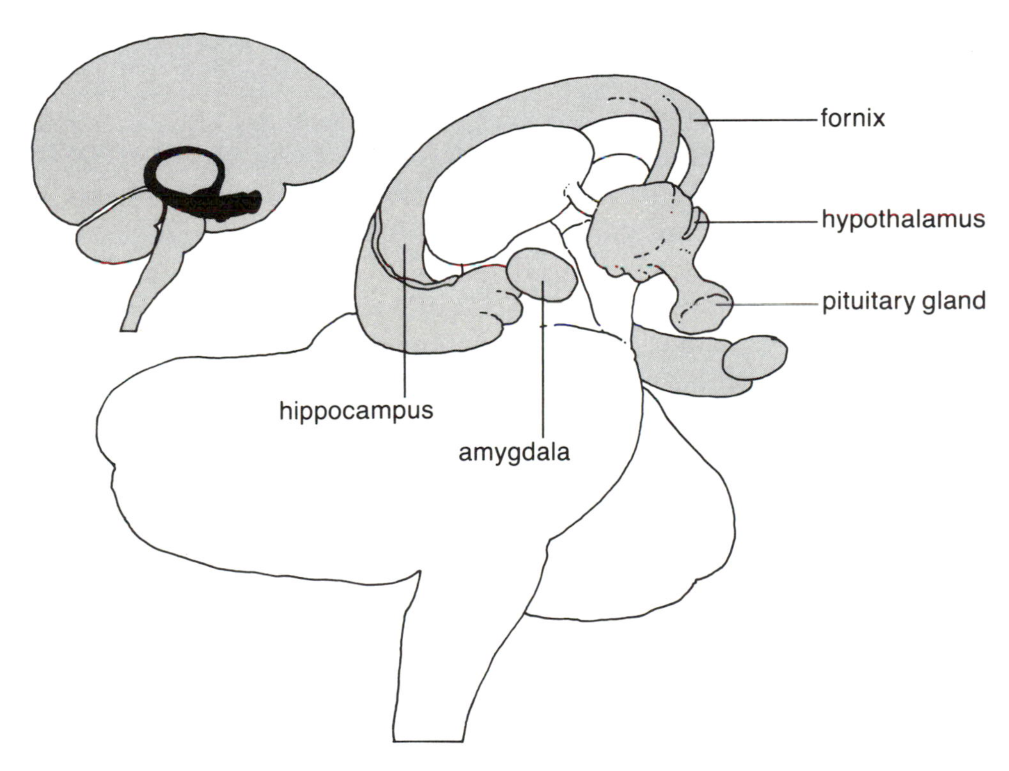

The Limbic System and Its Position in the Brain

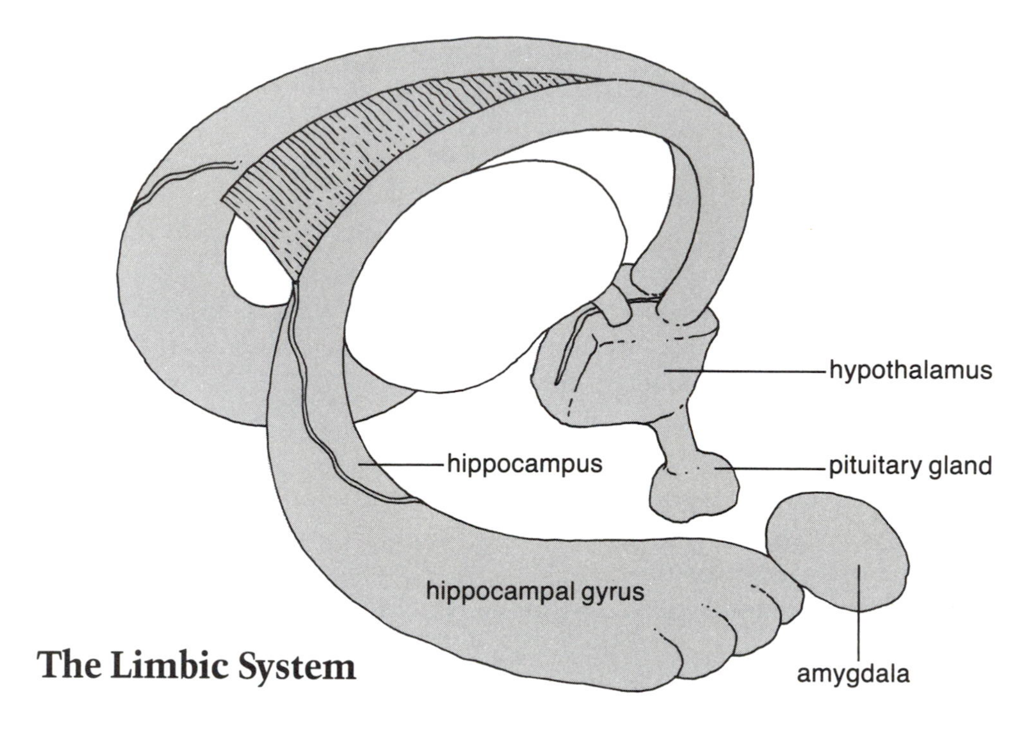

The Limbic System

in the emotional reactions that have to do with survival, such as sexual desire or self-protection through fighting or escaping. One way to remember limbic functions is that they are the "four f's" of survival: feeding, fighting, fleeing, and sexual reproduction.

The hypothalamus is perhaps the most important part of the limbic system. It is the "brain" of the brain. Without question, it is the single most intricate and amazing part of the brain. It is tiny, about the size of a pea, and weighs about four grams. It regulates eating, drinking, sleeping, waking, body temperature, chemical balances, heart rate, hormones, sex, emotions. The hypothalamus controls these homeostatic mechanisms of the body through feedback. For example, one's body temperature is monitored in the hypothalamus, using blood temperature as the control. If the blood becomes too cool, the hypothalamus reacts by stimulating the body's heat production and conservation process.

Through a combination of electrical and chemical messages, the hypothalamus also directs the master gland of the brain, the pituitary. This gland regulates the body through hormones. Hormones are chemicals manufactured and secreted by special neurons in the brain; they are then carried in the blood to specific "target cells" in the body. For example, in the male the gonadotropic (literally "toward the gonads") hormone is secreted by the pituitary and is carried to the testes by the bloodstream, where it stimulates the production of testosterone. This is the primary male hormone involved in both sex and aggressive behavior. The pituitary synthesizes most of the hormones used by the brain to communicate with the major glands of the body.

Two other major structures of the limbic system are the hippocampus ("sea horse"; it looks like one) and the amygdala ("almond"; it's shaped like one).

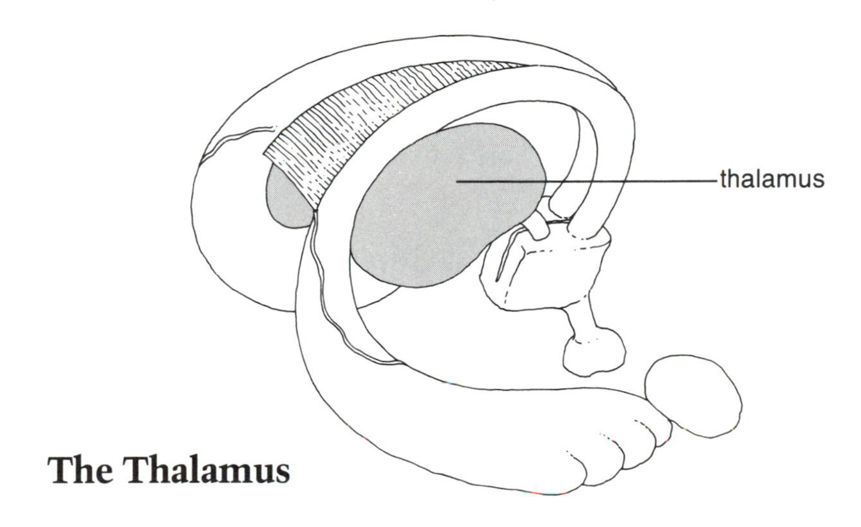

The Thalamus

The thalamus, located more or less in the center of the forebrain, helps initiate consciousness and make preliminary classifications of external information. Certain areas of the thalamus are specialized to receive particular kinds of information, which they then relay to various areas of the cortex.

Located on both sides of the limbic system, in each hemisphere, are the basal ganglia. Like the cerebellum, they are concerned with movement control, particularly with starting or initiating movements. In the human brain, these exquisite networks of cells are large and well developed. Although they are functionally rather different, the basal ganglia and the major structures of the limbic system are next to one another because they are both closely interconnected with the highest level of the brain—the cerebral cortex.

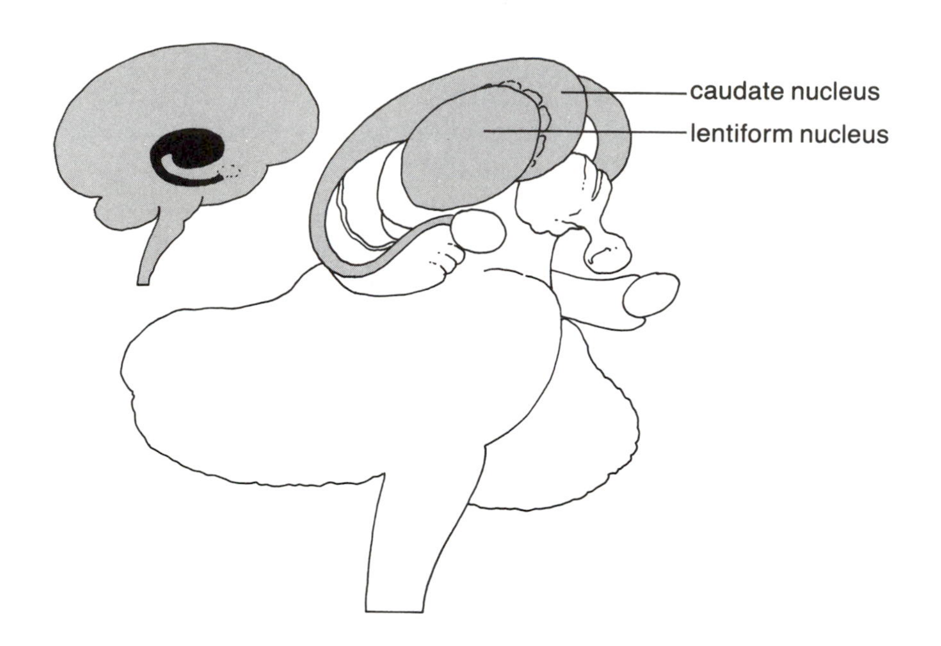

The Basal Ganglia

The surface of the cerebrum, which is made up of more neurons than any other brain structure, is called the cortex. It performs the functions that have greatly increased our adaptability. In the cortex decisions are made; the world is organized; our individual experiences are stored in memory; speech is produced and understood; paintings are seen; music is heard.

The cortex is only about one-eighth of an inch thick, but it is intricately folded. Of all mammals, humans have the most enfolded cortex, perhaps because such a large cortex had to fit into a small head to survive birth.

In what you are about to read it will seem as though a lot is known about the cortex, but in fact we know very little about how it works. We do know that certain activities are centered in the cortex, and we also know that some kinds of memory are cortical. We don't know, yet, exactly where memory is stored or how (although we have some clues), and we don't know how we "retrieve" specific memories. We do know that thinking and some aspects of learning are cortical functions, but we don't know exactly how we "get" new ideas or what happens in the brain when we learn something new. Study of the higher

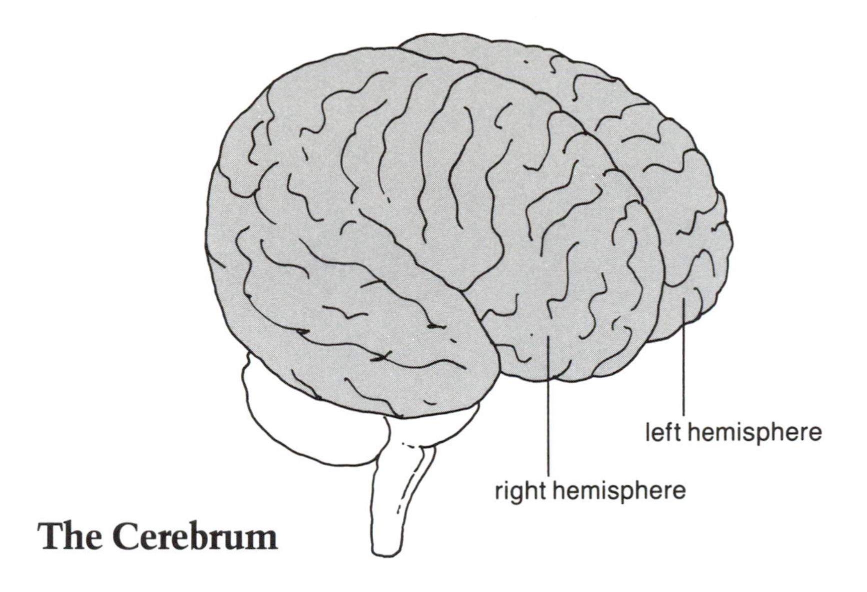

The Cerebrum

brain functions of the cortex is, and probably always will be, the frontier of research in the neurosciences, and applying all our marvelous cortical abilities to unraveling the mysteries of their operations is a challenging, and maybe impossible, task.

The cortex is the "executive branch" of the brain, responsible for making decisions and judgments on all the information coming into it

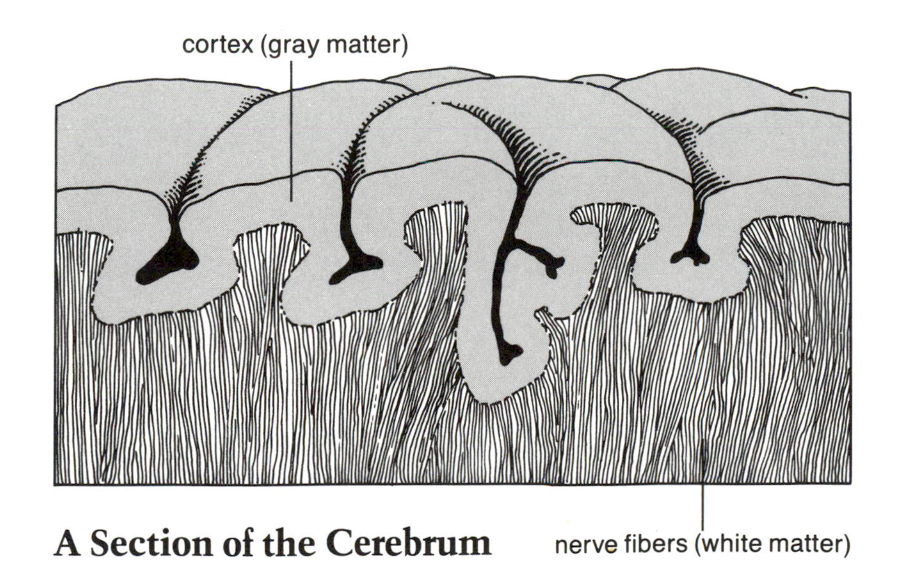

A Section of the Cerebrum

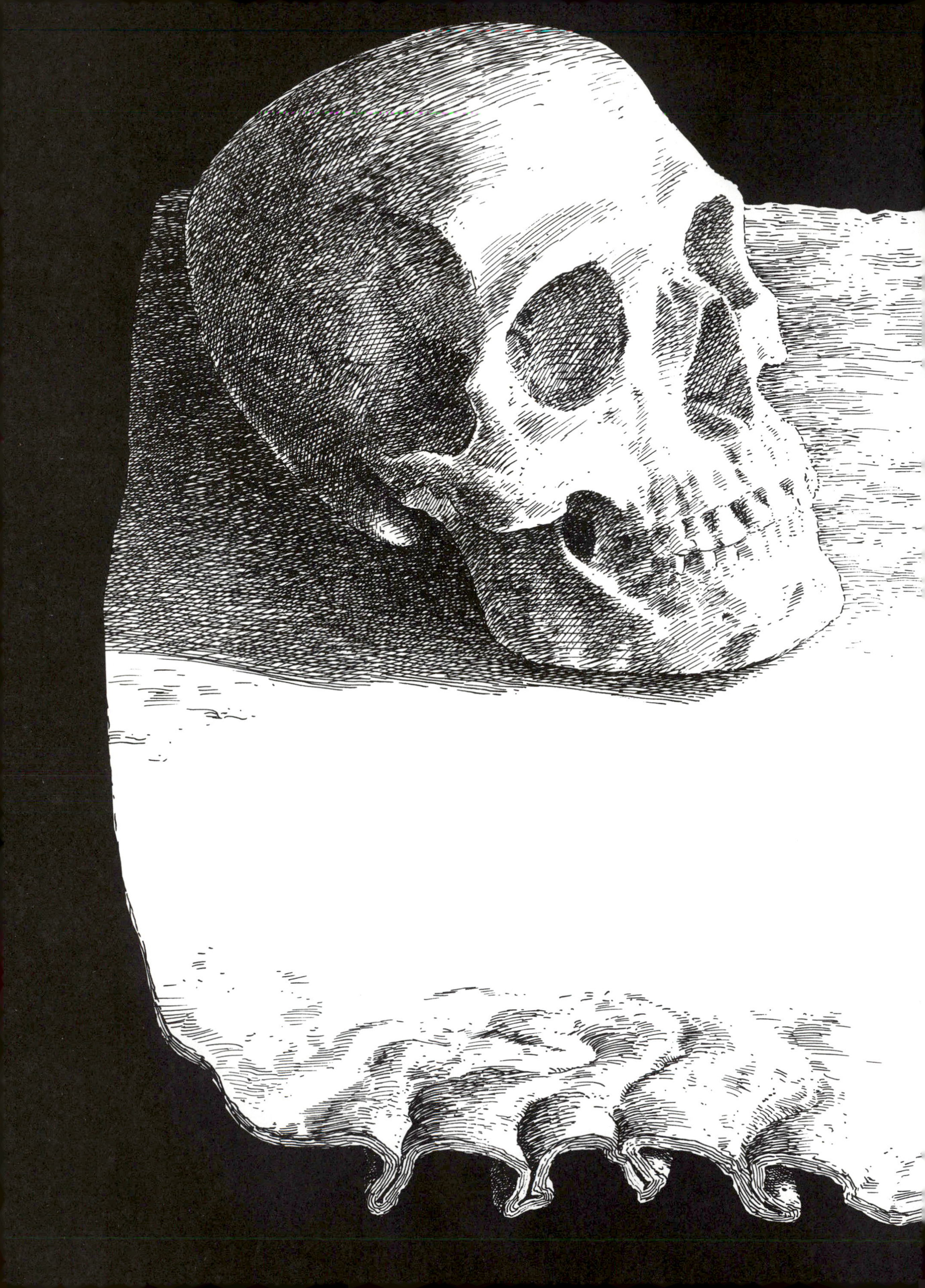

This illustration shows the approximate size of the unfolded cortex compared to the skull into which it must fit.

from the body and the outside world. First, it receives information; it analyzes and compares this new information with stored information of prior experiences and knowledge, and makes a decision; it then sends its own messages and instructions out to the appropriate muscles and glands.

The cerebrum, as we have seen, is divided into two hemispheres. Each one is responsible for the opposite half of the body. The left side of the brain controls movements and receives information from the right side of the body; the right side of the brain, from the left side of the body.

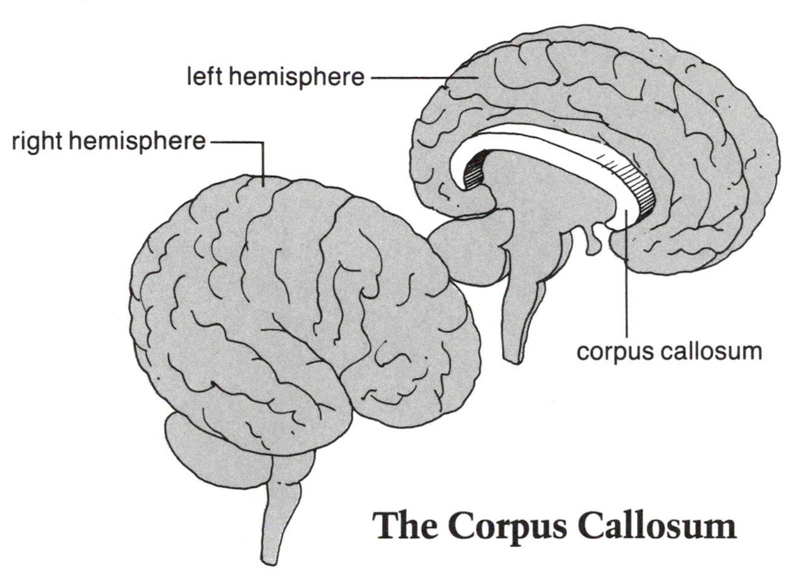

The Corpus Callosum

The two hemispheres are connected by a band of nerve fibers called the corpus callosum, the largest fiber pathway in the brain—a "bridge" of some three hundred million nerve fibers. Between four million and one million years ago, the fourth (and to date final) level of human brain organization emerged: the lateral specialization of the two hemispheres. These differences in function appeared at the time humans first began to make and use symbols (both language and art). One commentator has named this level of brain organization the asymmetric-symbolic level.

Each hemisphere is similarly divided into four areas called lobes. They are the occipital lobe, temporal lobe, frontal lobe, and parietal lobe.

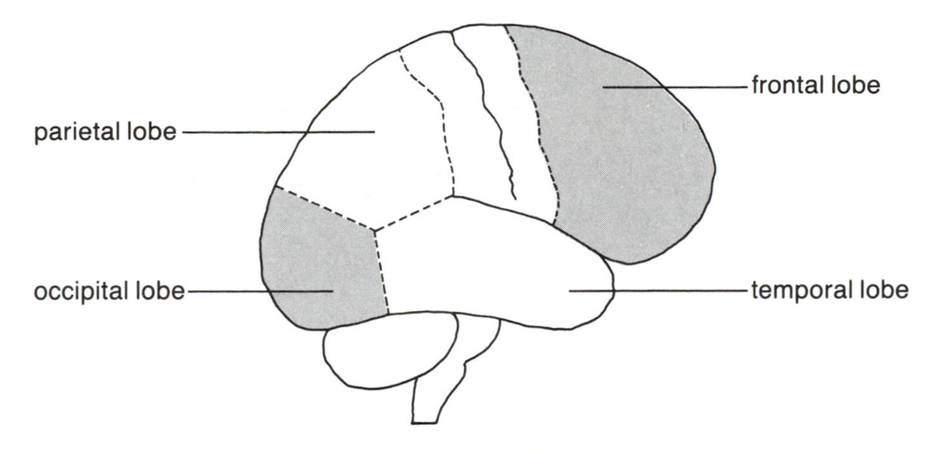

The Lobes of the Cortex

At the rear of each hemisphere is the occipital lobe. Because this area is devoted entirely to vision, it is often called the visual cortex. Visual information is sent from the eyes to the visual cortex and is analyzed for orientation, position, and movement. Damage to the occipital lobes can result in blindness even if the rest of the visual system is unaffected.

The temporal lobes (near the temples) have several important functions. In both hemispheres a part of the temporal cortex about the size of a poker chip is responsible for hearing; it is called the auditory cortex. Other temporal lobe functions appear to involve perception and memory.

Most of our knowledge of the workings of the temporal lobes comes from people who have suffered some sort of damage to this region. In some cases dramatic hallucinations occur; in other cases events that take place after the damage cannot be remembered. Severe damage to certain areas of the left temporal lobe, such as that caused by a stroke, may result in aphasia, the loss of language. Consider, for example, the case of one patient interviewed while in the hospital. His interviewer asks: "Can you tell me what work you have been doing?" He answers: "If you had said that, poomer, near the fortunate, tampoo all around the fourth of martz. Oh, I get all confused!" Notice that the phrasing and sounds of the language are like English. The patter is not completely random: it sounds like someone speaking unintelligibly in the next room. What is lacking is the ability to organize the basic sounds of language into something meaningful.

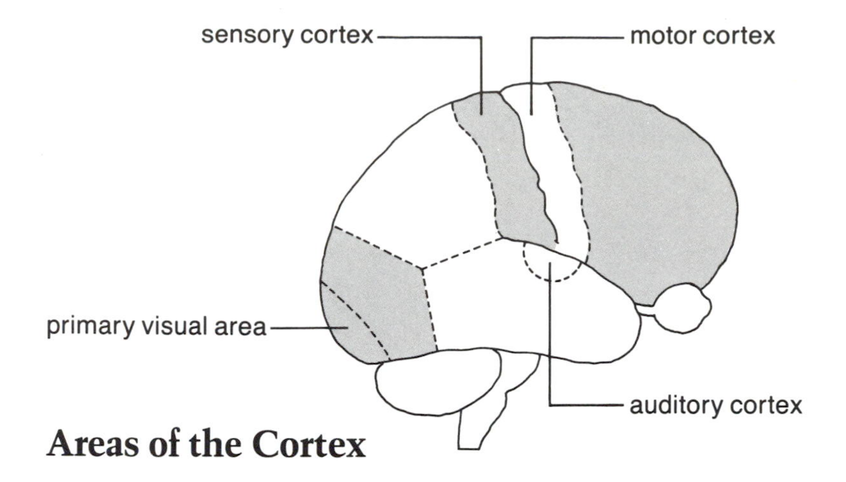

Areas of the Cortex

Damage to the right temporal lobe, on the other hand, results in impaired performance of spatial tasks, such as the ability to draw.

When the temporal lobe is electrically stimulated, some people report the feeling of being in two places at once: the memory of an event and the present *coexist* in the person's consciousness. Although fully conscious and aware of the operation going on, a person might suddenly feel he is also in a kitchen, thirty or forty years before: the sounds and smells seem real.

The frontal lobe, just behind the forehead, is the largest of the four cortical lobes and oversees much of the rest of the brain's activity. It has an especially rich connection with the limbic system. There is some evidence that an individual's initial appraisal of whether an event is threatening or dangerous is carried out in the frontal lobe. It is primarily involved in planning, decision making, and purposeful behavior. If the frontal lobes are destroyed or removed, the individual becomes incapable of planning, carrying out, or comprehending a complex action or idea, and unable to adapt to new situations. Such people are unable to focus attention and are extremely distracted by irrelevant stimuli. Although many of the most complex functions, such as language and consciousness, seem unimpaired, the loss of the ability to adapt and plan ahead makes all those other abilities less useful.

The parietal lobes (from the Latin for "forming the sides"), toward the rear of each hemisphere, seem to be where we assemble our world. It is probably here that letters come together as words, and words get put together in thoughts.

Damage to either parietal lobe can result in a form of agnosia (not knowing). Vernon Mountcastle studied a person with parietal lobe damage who was unaware of one whole side of his body, a condition called amorphosynthesis. Because his right parietal lobe was damaged, he ignored or did not "know" the left side. Drawings done by Mountcastle's patient show all the numbers of the clock crowded into the right half. A person with damage to one of the parietal lobes may only dress and groom one side of the body. Some individuals lose the ability to follow audio or visual clues and cannot recognize familiar objects by touch.

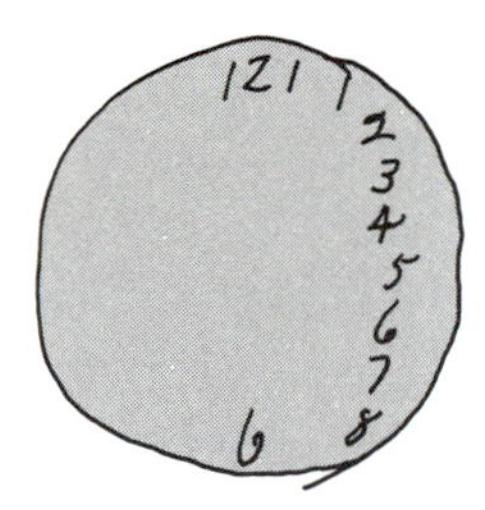

The somatic, or body, sensory areas are located near the parietal lobes. They are relatively smaller in humans than in other animals, but ony because the rest of the cortex is so large in humans. The sensory areas receive information about body position, muscles, touch, and pressure from all over the body, while motor areas control the movements of different parts of the body.

Information about the status and condition of different parts of the body is represented in corresponding parts of the brain, although the area in the brain devoted to one part of the body or another is not in direct proportion to the physical area of the body. The reason for this is that the more the function is used, the more space is given to it in the brain. Although the back is larger than the tongue, it makes fewer intricate movements and is less sensitive. Our hands are terribly important to us, giving us information on touch and pressure, and they are also capable of extremely complex movements. In a cat's brain, very little space is devoted to its paws, which provide little sensory information, but a very large area is devoted to the whiskers, which are far more sensitive. In the rat, specific cells actually respond to specific whiskers. We'll discuss this in more detail in the next chapter.

The cerebrum's two hemispheres look almost identical, but there are significant anatomical differences. An examination of fetuses and stillborn infants reveals that in 95 percent of the cases a part of the left hemisphere is larger than the right. The enlarged area in the temporal lobe is called the planum temporale, which is involved in speech and written language.

Although each hemisphere is specialized to handle different tasks, the division between them is not absolute—they are in constant communication with each other. Rarely is one hemisphere completely idle and the other frantic with activity. The left hemisphere is much more involved and more proficient in language and logic than the right; the right is much more involved in spatial abilities and "gestalt" thinking than the left. But it is oversimplified and misleading, even wrong, to assume that the two hemispheres are *separate* systems, "two brains." An activity as complex as language involves both hemispheres interacting with each other. If either hemisphere is damaged, the remaining "intact" hemisphere can take over, but this becomes less easy as we age. If the left hemisphere is damaged at birth, the right will take over language, although the person may be less adept at language than he or she otherwise would have been.

The recent (in evolutionary time) division of the functions of two hemispheres of the brain is most distinctively human. We have language, art, inspiration—kept separate by the enormous chasms between two brain structures, and kept in touch by an enormous spanning bridge. It does make you wonder how all this happened.

The human brain's most celebrated achievement (by us humans!) is thought. Those things that are thought to make us most human—language, thinking, perception, intelligence, consciousness—represent only a small fraction of the brain's functions. Scientists make a serious error in confusing what is *uniquely* human with what is *typically* human. What the brain chiefly does is regulate the body. It controls body temperature, blood flow, and digestion; it monitors every sensation, each breath and heartbeat, every blink and swallow. Much of its work is in directing movement—walk this way, take the hand off the stove, lift the arm to catch the ball, smile. Even speech is movement—the tongue, the lungs, the mouth, and the pharynx all must be directed to move in certain ways to produce speech.

To follow the progress of the human brain, one must remember that it is designed primarily to run the human body.

There is a footprint in Africa, impressed in the sand more than three and a half million years ago. It marks a spot where human beings diverged from the rest of creation. It is the footprint of a creature who is beginning to stand up on two legs, not four. That first tentative step set into motion a series of evolutionary "steps" that made modern humans who we are. The shift from four legs to two not only caused our ancestors to rely less on smell and more on vision, it also freed the front limbs for other activities, such as toolmaking and carrying. This contributed to the development of language and, ultimately, of modern society.

Here's how it works from the architect's viewpoint: with the freeing of the front limbs, the hind limbs have to bear the entire weight of the body. The human back was not originally "designed" to support upright posture (which partially explains why back pains are a common complaint). To support the additional weight, the human pelvis grew thicker than that of the great apes. The thickened pelvis made the birth canal, the opening through which infants are born, much smaller.

But while the birth canal was becoming smaller, the brain and head were growing larger. If there had been no correction for this new disadvantage, the human species eventually would have died out because of inefficient childbirth. It seems that the "solution" was to have human babies born *very early* in their development. At birth a chimp's brain is 45 to 50 percent of its adult weight, but a human baby's brain is only 25 percent of its adult weight. Human children have the longest period of helplessness in the animal kingdom. So the major portion of the brain's development occurs outside the womb; it is exposed to and influenced by many different environments, experiences, and people.

The brain evolved faster than any other organ in history: it took hundreds of million of years to create the 400 cc brain of *Austalopithecus* four million years ago in Africa, yet in only a few million more years the brain had grown to 1250 to 1500 cc, and had developed the capacity for abstract thought. It has helped us to adapt to every kind of geography and climate.

Our brain is the largest, relative to body size, of all land mammals, but it is not just the size of the brain that matters. What is especially important is *where* the brain is large. Our cerebral cortex, the uppermost part of the brain, is much larger and more intricate than in any other animal. It is the most distinctive part of being human. It enables us to carry ourselves beyond our inheritance, and to create our own environment—again and again.

This is how, over millions of years, our brain got to be the way it is, a ramshackle structure, with elements of our reptile and mammal past, and with an ability (we hope!) to create our own future. In this chapter we've considered the *overall* structure of the brain. The next three chapters consider the *specific* structures of the brain. First, we take a look at how the upper "rooms" of the brain are built.

2

The Sensory Brain: The Columns of Experience

THE BRAIN, as we have said, is like a ramshackle house with many different "rooms." Some of the most important rooms have to do with our experience of the outside world. You open your eyes and see a beautiful three-dimensional view of forms and colors. But it is much more than that—it is highly organized and structured. This form and structure of the world you see is created by the visual part of the brain, particularly the visual area of the cerebral cortex, the sheet of cortical tissue at the back of the brain that codes the visual world.

We also experience the world through hearing and touching. There is an auditory area of the cerebral cortex as well as one devoted to the sense of touch. So far as we know, all the sensory areas, or fields, on the cerebral cortex work in fundamentally similar ways to code the vast and complex array of stimuli into our highly organized and sensible experience of the world. Although our oldest and most basic senses are taste and smell, their mechanisms are much more difficult to explore, and we can only assume that they operate in ways similar to our senses of touch and sight.

Because we know more about the brain systems that code seeing than about other sensory systems, we will focus here on the visual system, and particularly on the visual area of the cerebral cortex, as the best-understood part of the sensory brain. A most remarkable discovery about this sheet of cortical tissue was made a few years ago by David Hubel and Torsten Wiesel at Harvard, for which they were awarded the Nobel Prize in 1981. The visual cortex is organized into many

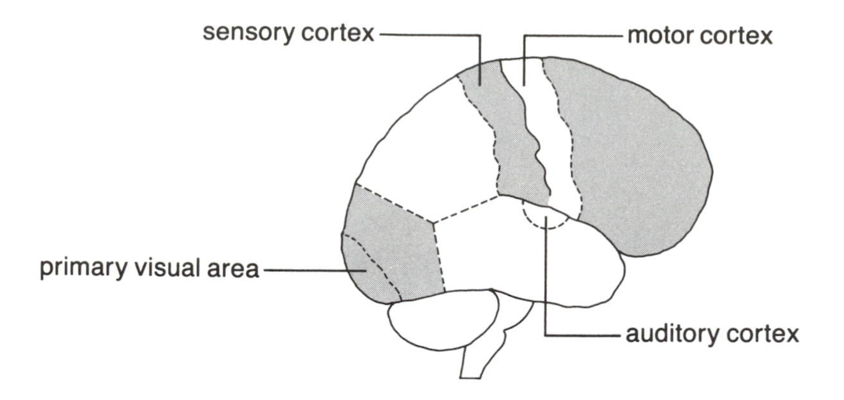

thousands of little columns of nerve cells that run from top to bottom through it.

These little columns of neurons seem to code the basic aspects of visual experience. The same is true for the area of the cerebral cortex that serves the sense of touch — it is made up of thousands of little columns that code the basic aspects of skin sensations, such as touch and pressure. We think the same is true of the cortical area that codes hearing, although less is known about it.

So, as we step into the visual "room" — the visual area of the cerebral cortex — we find that it is filled with thousands of columns of nerve cells. In fact, there are several visual rooms. We can then move down from the large visual room to see how it is organized as several rooms, each filled with many little columns of nerve cells that are truly the columns of experience.

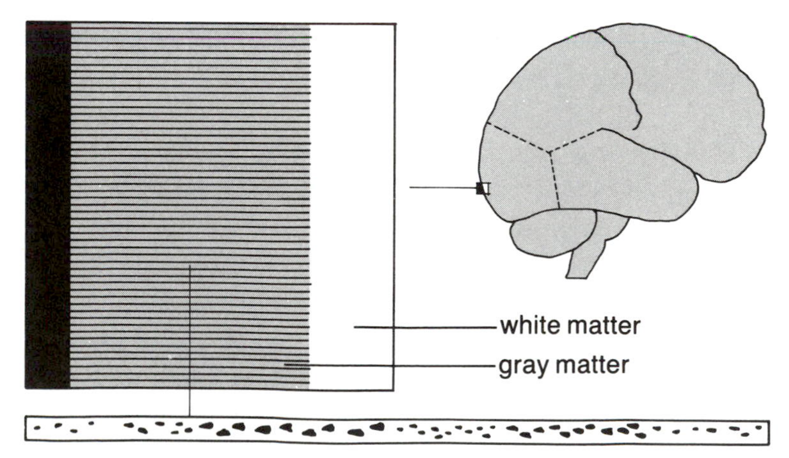

Columnar Structure of the Visual Cortex

First, let's consider the story of "Bobby": When Bobby was two years old he developed a small growth on his left eyelid. It was not really a serious matter at all — the growth was not malignant. But Bobby worried it and picked at it until it became infected. The doctor treated it and covered Bobby's left eye with a bandage so Bobby could not get at it. After a week, the bandage was removed and the eyelid was fine. Indeed, Bobby's parents forgot all about it.

In first grade Bobby's eyes were tested in the school program. Bobby's right eye had normal vision, but his left eye was poor. Bobby was taken to the ophthalmologist to have his eye refracted. Using special tests, the ophthalmologist can determine what is wrong with the eye.

It seemed very likely that Bobby was nearsighted in his left eye. This meant glasses, a small price to pay for good eyesight. To everyone's astonishment, there was nothing at all wrong with Bobby's left eye. The lens was perfect — it projected the sharp image of the world exactly on Bobby's retina. The retina was also completely normal and was working perfectly. Yet Bobby had poor vision in his left eye. The ophthalmologist could not explain it and was unable to treat it. Bobby had poor vision in his left eye throughout life. No one thought it could be due to the eyelid infection he had had at age two, and indeed it was not directly due to this. The infection never went beyond the eyelid and had no effect on the eye.

We now know why Bobby had poor vision in his left eye. Believe it or not, it is because his left eye was kept closed for a week when he was two. We'll see why this is so later in this chapter. This new understanding is the result of a fascinating scientific detective story — the story of how we see, how the eye and the visual part of the brain work, and how they form the architecture of seeing. (Although Bobby's story is fictional, it is based on our current understanding of the mammalian visual system and some clinical evidence.)

The eye works very much like a modern 35 mm camera. It has a very high quality lens made of translucent cells. In fact, the lens of our eye is better than any glass lens yet developed for cameras because it can change its shape to focus on distant or near objects. In a camera the lens has to be moved forward or back to do the same. In a normal eye, the lens focuses a sharp image of the world on the retina at the back of the eye, just as in a good camera.

The lens projects a crisp and clear image of the world on the retina. This activates the photographic "film" in the retina. From this point

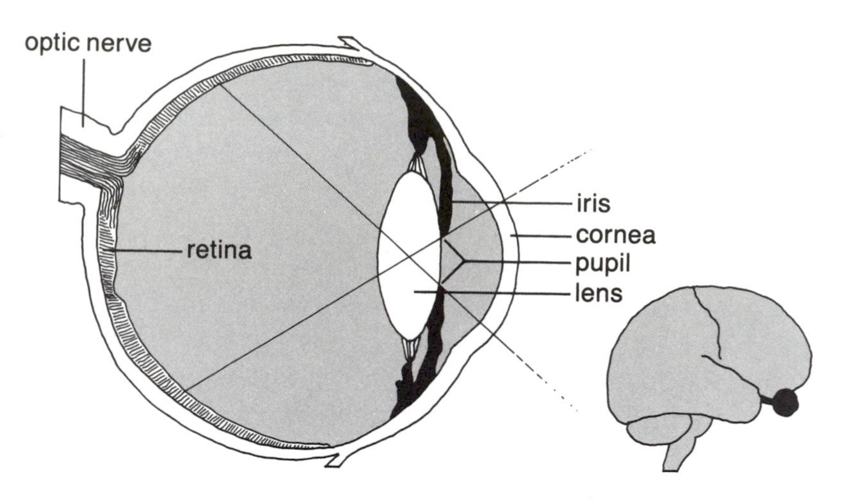

Cross Section of the Eye

on, the eye differs very much from a camera. The film in the eye is a layer of light-sensitive cells, the rods and cones. They contain visual pigments—colored chemicals that react to light in a manner analogous to the way grains of silver in a photographic film react to light. But the pigments in the eye do not change permanently. Instead, they change temporarily to a degree, depending on the amount of light. When the light is removed they return to their original state. The pigments of the eye are composed of chemicals made from vitamin A and proteins, which explains why carrots, high in vitamin A, are good for your eyesight.

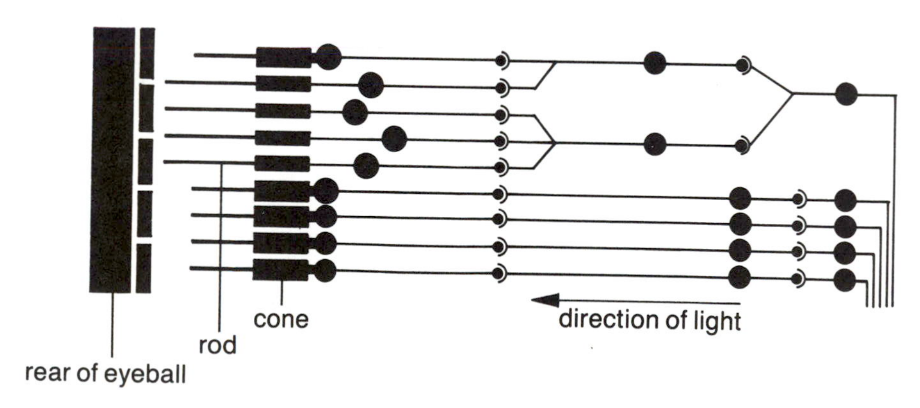

Schematic Diagram of the Retina

The picture of the world seen by the retina of your eye is very different from the picture of the world you actually "see." The retina sees the world as a series of spots or dots of light, dark, and color. When you look at the world, your eye conveys these dot patterns to your brain, which converts these dot images into the smooth and continuous "seeing" that you experience. To understand the architecture of seeing, we must first understand the architecture of the visual cortex.

The whole cerebral cortex contains billions of nerve cells, more than all the rest of the brain put together. Yet its structure has an elegant simplicity. There are six layers of nerve cells, each layer containing cells with certain shapes that interconnect with one another both within and across layers (for purposes of discussion, these layers are numbered I through VI). Think of the cortex as a six-layered sheet laid out to cover the brain.

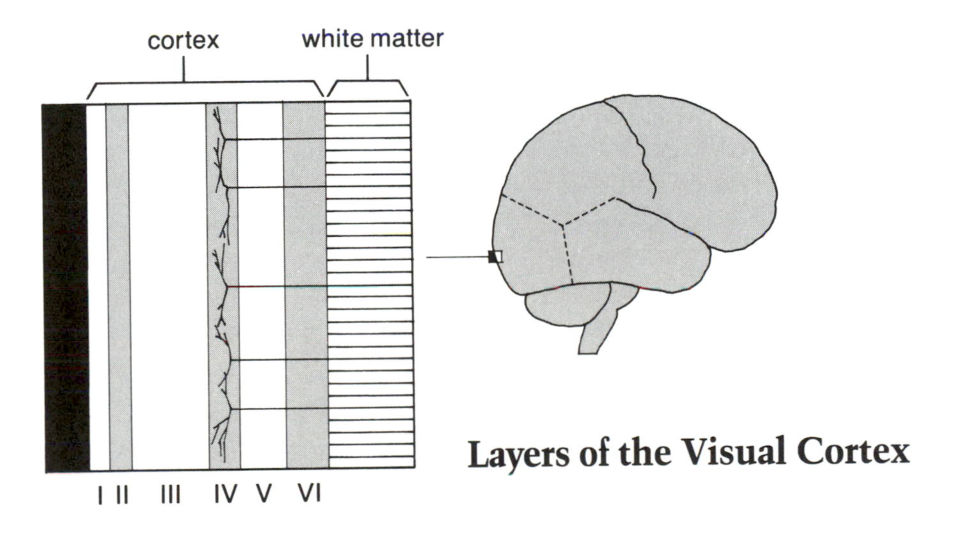

Layers of the Visual Cortex

There is a second aspect to the structure of the cortex: it is organized into columns of cells from the upper surface down, running through the six-layered sheet. The many cells from the top to the bottom are interconnected with one another to form structural columns. These vertical columns are not very big, a fraction of a millimeter in diameter, and seem to be present everywhere in the cortex. Many neuroscientists now feel that the column is the basic functional unit of the cerebral cortex.

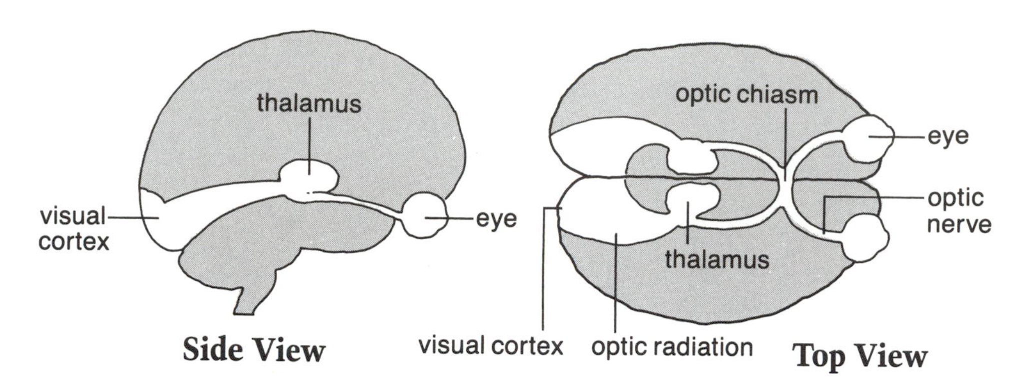

The Primary Visual Pathway

Nerve cells carrying information from the eye terminate on cells in one particular layer of the visual cortex, layer IV. (Actually, the nerve fibers from the eye connect to nerve cells in a lower region of the brain —a way station in the thalamus—and these cells in turn connect to cells in layer IV of the visual cortex. However, the dot patterns from the eye are presented unchanged to the cells of layer IV in the visual cortex, so we can ignore the way station.) The dot picture from the retina is projected to the visual cortex over its whole extent. If we were to stretch the visual cortex into a sheet, it would of course be many times larger than the size of the retina, and though it receives a very precise picture from the eye, the picture is distorted.

When you look at an object, the center of your gaze is projected to a tiny region of the retina, about one millimeter on a side, where you see details most clearly. This region, in turn, is projected to half the entire visual cortex. Much larger regions to the sides of the retina are projected to smaller areas of the visual cortex. The reason you see details best at the center of gaze is because so many cells in the visual cortex are acted on by this small center region of the retina, and many more neurons in the cortex are devoted to processing information from this small part of the retina.

Imagine that you are a nerve cell in layer IV of the visual cortex. What would you see? What kind of information would act on you? In essence, you will see a dot of light, one elementary spot—but only if the small spot of light falls on the small part of the retina in the eye to which you are connected. If light does not fall on that spot on the retina, you, the nerve cell, will not see anything. Only one little region on the retina is the receptive field for a particular nerve cell in layer IV of the visual cortex. The nerve fibers from this little area of the retina, and only from this area, connect to a given nerve cell in layer IV. When and only when light falls on this receptive field in the retina will the cell in layer IV of the cortex be excited. This receptive field can only be in one eye, either the left or the right.

If we move one step away in layer IV of the visual cortex and look at a neighboring cell, it will have a receptive field in the other eye but in the same location on the retina of that eye. There are adjacent columns of cells in layer IV, so that if one group has a receptive field in the left eye, the next column will respond to light on the same place in the right eye. These groups of cells alternate along layer IV throughout the visual cortex. Suppose we shear off the top three layers of the cortex and spread out layer IV. Suppose further that we label the right-eye input as dark and the left-eye input as light. We would see a regular pattern of stripes. These stripes are the inputs from the two eyes to the cells in layer IV of the visual cortex.

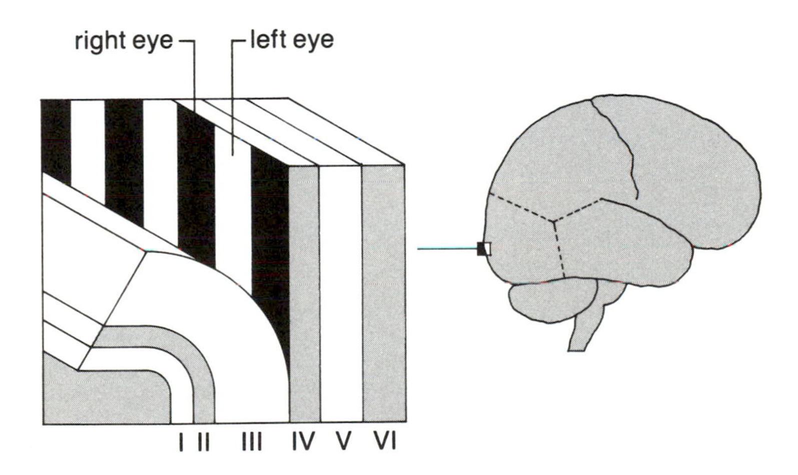

Ocular Dominance Areas of Layer IV

What about Bobby? His left eye functioned normally but had poor vision because something went wrong with his visual cortex. Studies on animals provided the answer. The visual cortex of higher mammals like the cat and monkey is essentially the same as the human visual cortex. The pattern of stripes in layer IV looks the same as in adult

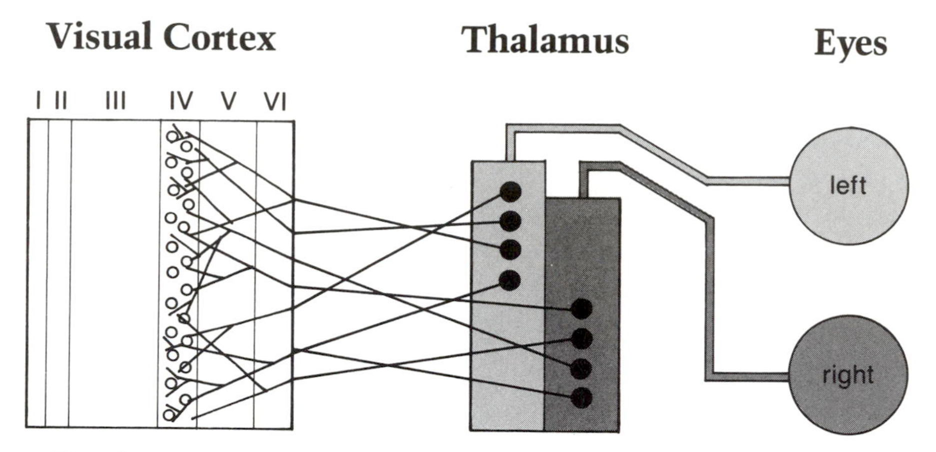

a. The visual system two weeks after birth (no ocular dominance in visual cortex)

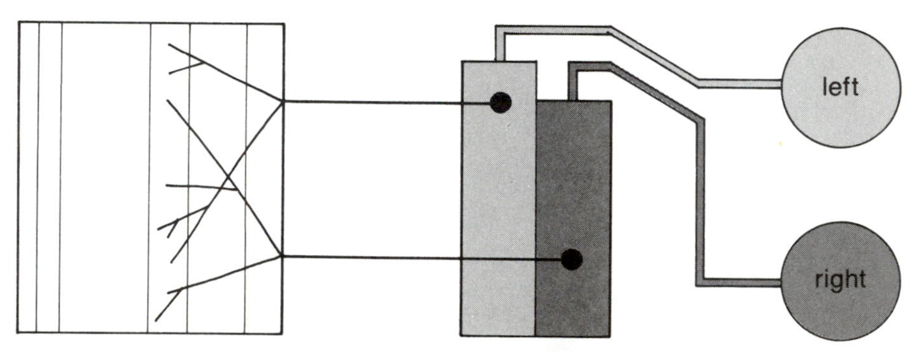

b. Development of adult pattern through competition

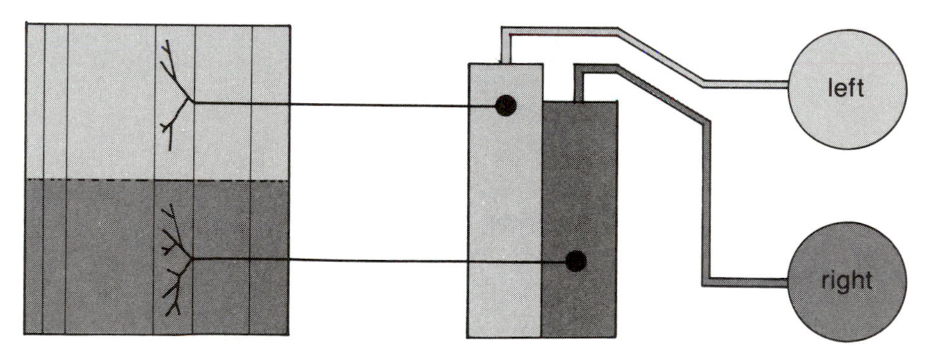

c. The strongest synaptic input dominates and takes over columns of cells, creating alternating areas of ocular dominance in layer IV of the adult visual cortex.

humans. But the pattern is very different in a newborn animal. Each eye projects to virtually all the cells in layer IV. During infancy the pattern shifts gradually to alternating stripes.

At birth the fibers from the two eyes act together on all the neurons in layer IV of the visual cortex. A light in either eye will activate a given nerve cell in layer IV. But soon the fibers coming from the two eyes compete. In one small region of the visual cortex the right eye will have a small advantage and will eventually win out. In another region the left eye will win. Throughout infancy the fibers from the two eyes constrict from complete overlap to complete separation in the adult pattern of stripes. We do not yet know how this occurs, but it does. It is perhaps an example at the cellular level of competition and the survival of the fittest.

If one eye is kept closed in an animal during infancy and the covering then removed, the animal will be completely and permanently blind in that eye. If the same is done in an adult animal, vision in that eye remains normal. Indeed, adult humans who develop cataracts can see quite normally with appropriate glasses after the cataracts are removed. Tragically, a human infant born with one cataract will be blind in that eye after the cataract is removed if the surgery is delayed very long.

In an animal raised through infancy with one eye closed, the fibers from the closed eye lose out almost completely to the fibers from the normal eye in their competition for cells in layer IV of the visual cortex. The normal eye comes to activate most cells in layer IV and the closed eye ends up activating almost none. This change seems to be permanent. The reason it occurs is simple enough. Normal visual stimulation is necessary for sight to develop. Visual experience allows the fibers from the eye to keep their competitive edge and establish their stripes of dominance in the visual cortex. Without this normal visual stimulation, they lose to the other eye.

The critical period during which the two eyes establish their zones of dominance seems to be about the first six years in humans, six months in monkeys, and perhaps three months in cats. It is a very sensitive period. If one eye of a kitten is kept closed for only one day, it will have poor vision in that eye as an adult!

There is a very important practical lesson from this basic work on the visual brain. Do not ever keep *one* eye of a human infant closed

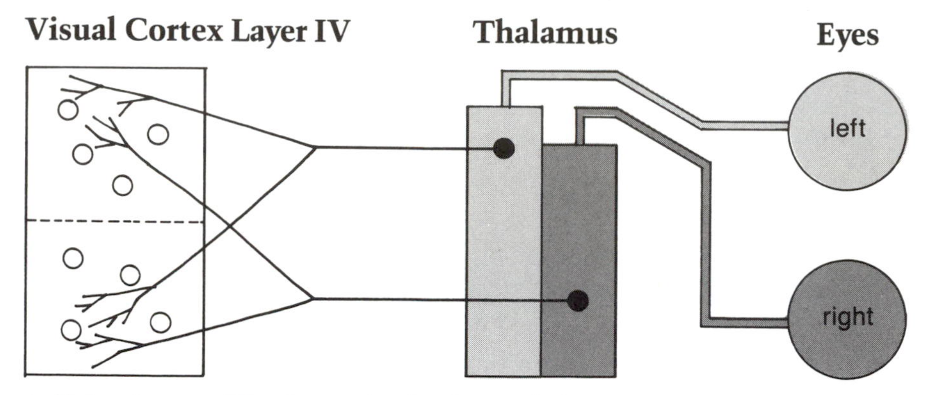

a. Natural competition for ocular dominance between the left eye and the right eye

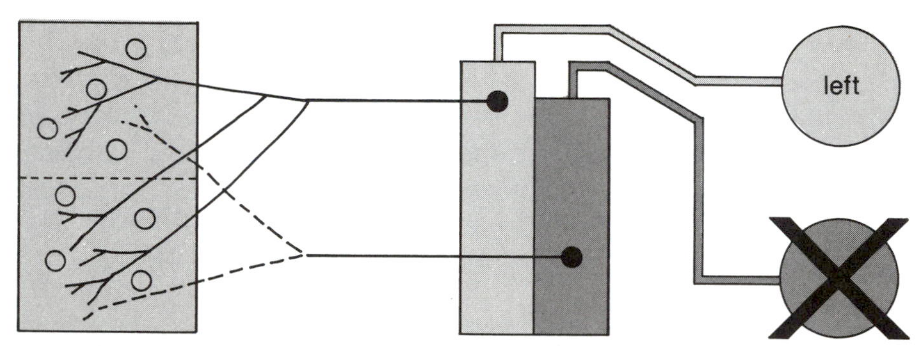

b. With the right eye closed, synaptic connections triggered from that eye stop

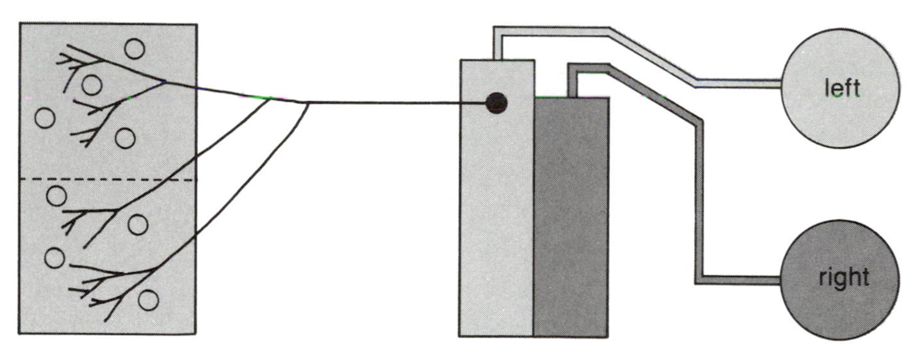

c. The left eye now completely dominates both areas of layer IV

Effects of Closing One Eye on the Development of the Visual Cortex

for an extended period of time. Keeping *both* eyes closed is better; after all, infants sleep a good bit of the time. It is the competition between the two eyes for cells in layer IV of the visual cortex that determines how we see, and in infancy that competition is fierce.

Any cell in layer IV receives information from only one eye or the other, and, in fact, all it "sees" is dots of light. The receptive field in the retina for a cell in layer IV of the visual cortex is the circular area of the rod and cone receptor cells that transmit information from the light that falls on them. Actually, the receptive fields on the retina are such that a neuron in layer IV responds most if a visual stimulus falls on *most* of its field rather than if the stimulus falls on *all* the field. Cells in the other layers of the visual cortex are influenced to varying degrees by the cells in layer IV. It is here that the architecture of the visual cortex converts the dot patterns from the eyes into the smooth and unified experience that is our visual world, or so we think.

Cells in the other layers of the visual cortex are informed by the layer IV cells of both eyes. However, a given cell may favor one eye much more than the other. So the columns of cortical cells appear, because any given small column of cells through the visual cortex from the top down has cells that respond much more to one eye than the other. If the layer IV cells in a column respond to the right eye, the cells in the other layers in that column will respond more to the right eye. Cells in the next column will respond much more to the left eye, and so on.

An extraordinary part of the story comes next. The cells in layer IV "see" dots of light. But the cells in the other layers "see" lines and edges. When you look at a line or edge, say the edge of a table, it will be projected along the retina of the eye. This will activate a row of cells in layer IV of the cortex. Each cell responds to the small dot of light that is the part of the table edge falling on its receptive field in the retina. A cell in layer V is activated by the row of layer IV dot cells that are stimulated by the table edge only if the appropriate row of layer IV cells is activated. The cell in layer V, however, "sees" an edge and not a dot of light. The row of light spots has been converted into a line.

Furthermore, the cell only sees the edge of the table if it has a particular orientation, perhaps horizontal. A nearby cell will only see the edge if it is tilted so many degrees away from horizontal. It is here

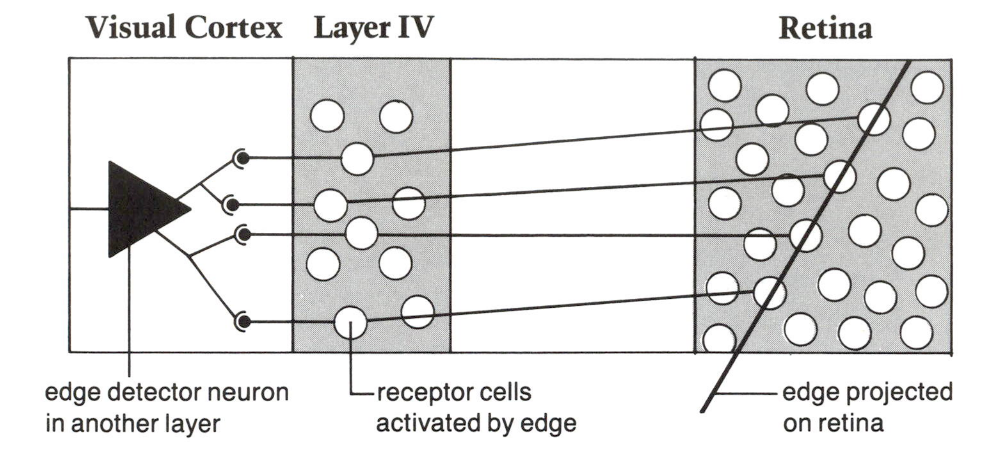

that the columnar architecture of the visual cortex is most striking. Consider a column of cells that is dominated by the right eye. It is made up in turn of many smaller columns, each tuned to a particular orientation of lines. One tiny column of cells will respond best to horizontal lines, the next to a line a few degrees from horizontal, and so on.

At this point the story begins to get even more complicated. The hard data — the facts — are mostly descriptions of how various nerve cells in the visual cortex respond to visual forms, but it seems reasonable to suppose from the examples that we have discovered that the ways in which these nerve cells respond to such stimuli form the basis of our visual experience — how the world looks to us. For example, some neurons respond best to two edges of lines joined to form a right

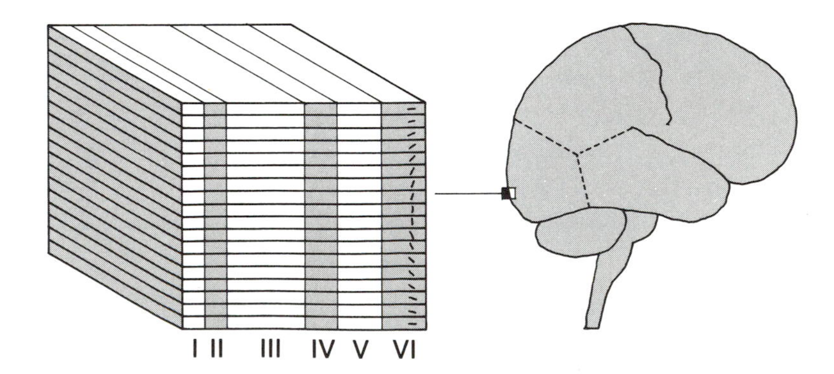

Schematic View of Orientation Slabs in the Visual Cortex

angle. Such neurons have "right angle" receptive fields. We often say loosely that they code "right angle." We don't really know exactly what such neurons are coding; we are simply making an educated guess from the kinds of visual stimuli the neurons respond best to.

The edge detectors, the neurons that respond to an edge at a particular orientation, are said to have "simple" receptive fields. It is easy to see how a row of dot cells in layer IV could connect to the edge detector cell so that it would only be activated by an edge of line of a particular orientation, an edge falling on a particular row of receptor cells in the eye. That is to say, any given simple edge detector neuron will only respond if the edge strikes a particular exact location on the retina of the eye.

Other types of neurons found in the visual cortex are more complex than the edge detectors or the angle detectors and seem to respond to particular sizes and forms of objects over wider regions of the retina.

So far, we have spoken only of the primary visual area, the region of the visual cortex that receives input from the eyes. This area on the cerebral cortex has a complete map of the surface of the retina, the receptive surface of the eye. (Actually, the visual area on the left visual cortex would have a complete map of the left half of each retina and vice versa for the right area, but we can ignore this detail for now.) A complete map is just that—if a small light is shined on each different part of the retina, a corresponding part of the visual area will respond.

If we move along the surface of this primary visual cortex, heading toward the front of the brain, we would next encounter another visual area that also has a complete map of the retina laid out on it. However, this second visual area receives much of its information from the primary visual area rather than from the eye. As we continue our journey forward, we cross the second visual area and enter a third one, and so on. At last count there were perhaps fifteen visual areas in the cortex of the monkey and presumably as many or more in the human cerebral cortex. Each of the visual areas has the same basic organization as does the entire cerebral cortex—each is composed of six layers of nerve cells, a six-layered sheet, and each has many thousands of little columns extending down through the six layers.

For convenience we refer to these areas as V1, V2, and so forth. V1 was the original primary visual cortex described by the anatomists. The neurons we have discussed so far, those in layer IV that see spots, those

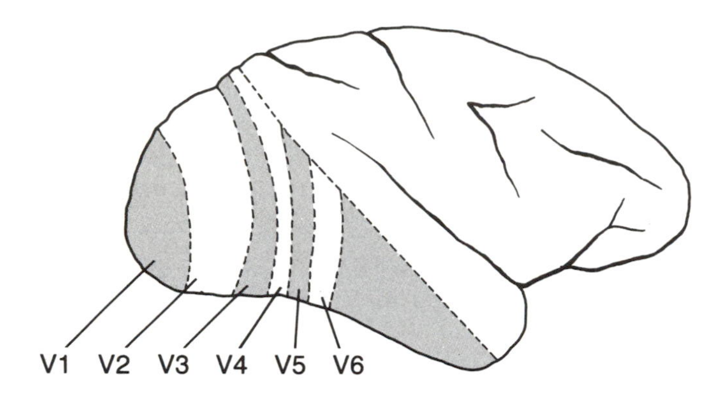

Areas of the Visual Cortex in the Monkey Brain

in other layers in the columns that respond to oriented edges and lines, much more from one eye than the other, and the more complex cells, are found in V1. V1 receives its visual input from the eye by way of the thalamus to the cells in layer IV. Most of the input to V2 is from V1, most of the input to V3 is from V1 and V2, and so on. So there is a progressive funneling of visual information from V1 to V2 to V3 to V4, and so on.

Another major difference is that most neurons in areas V2, V3, and on respond equally well to input from the two eyes, whereas in V1 any given cell will respond best to only one eye. In many of the cells in areas V2 and on, the response is a bit different for one eye versus the other eye. We see a three-dimensional world because we see it with both eyes. If an object is far away, the two eyes are almost parallel. As an object gets closer to the face, both eyes point more and more inward toward the nose. As a result, the image of the object falls over a slightly different region of the retinas of the two eyes. This retinal difference is noted in the visual cortex, and the disparity is the major reason that our visual world is in three dimensions. If you hold a finger upright in front of your nose and then shut one eye or the other, you will see it switch back and forth in relation to more distant objects.

Three-dimensional movies make use of the retinal difference by using two side-by-side cameras with different colored filters. When you view the movie through the cardboard glasses with a different colored filter over each eye, it suddenly becomes three-dimensional because each eye is seeing a different series of images (depending on the color of the appropriate filter), just as you normally see the world through both eyes. The fact that the binocular cells (cells that respond to input from both eyes) in the secondary visual areas respond somewhat differently to input from the two eyes reflects the retinal difference, or disparity cue. It seems very likely that these binocular cells are the neuronal basis of depth perception.

Why on earth should there be so many visual areas in the cerebral cortex? We are just beginning to get some ideas from current research. It seems that several of these secondary visual areas may play very special roles in particular aspects of seeing. One example is the perception of movement. Cells in all visual areas tend to respond better to moving objects than to objects at rest. However, cells in one of these areas respond particularly well to movement of an object across either eye. Neither the shape of the object nor the direction of movement matters, only movement itself. This visual area seems specialized to see movement.

Color is the most immediate visual sensation. Imagine describing what it is like to see to a person blind from birth. It is easy to describe the sizes and shapes of objects in words, but there is no way at all to describe colors. Color is "given" by experience. For instance, infants can match colors to samples long before they can learn the *names* of the colors. Early experiments on English-speaking adults showed that memory for colors that had been viewed is much better for the primary colors—red, yellow, blue—than for in-between colors that are harder to describe. At first this was attributed to cultural learning, since in the English-speaking world, the primary color names are set by the language. In fact, just the opposite is true. Anthropological studies of a people who have only two color terms in their language indicate that they remember all three primary colors best even though they do not have words to describe them all. Our immediate experience of colors is given by the color receptors in the eye, and we do not learn to see colors, we only learn the names our culture has given them.

"Color" as such does not exist in the world; it exists only in the eye and brain of the beholder. Objects reflect many different wavelengths of light, but these light waves themselves have no color. Animals developed color vision as a way of telling the difference between various wavelengths of light. The eye converts different ranges of wavelengths into colors, and it does this in a very simple way. There are two types of light-sensitive receptor cells in the retina. The rods see shades of gray and are more sensitive to dim light. In the human eye there are also three types of cone color receptors, having three different light-sensitive pigments — red, yellow-green, and blue. These connect to different nerve cells and thus send color information to the brain.

There are some neurons in area V1 that respond to color — more to one color than another — particularly in the region representing the center of gaze of the retina. This region of best detail vision in the eye contains only tightly packed cones and no rods. In general, however, color coding is not prominent in most visual areas. Instead, a whole separate visual area seems devoted to perceiving color. Most cells in this area are highly selective and individually respond only to a narrow range of wavelengths. Overall, different cells respond to all the wavelengths of light we see as colors. These cells are not particularly interested in the shape, size, or movement of stimuli, only in color. One cell was even found that responded best to magenta, a color that is outside the normal color spectrum and is produced by superimposing red and blue.

There is still another visual area that seems to specialize in seeing the shapes of objects. In monkeys, when this area is damaged, the ability to tell different two-dimensional patterns apart is lost. However, simpler aspects, such as size, can still be seen.

One final visual area is perhaps the most remarkable of all. It receives the complex and highly processed input from other secondary

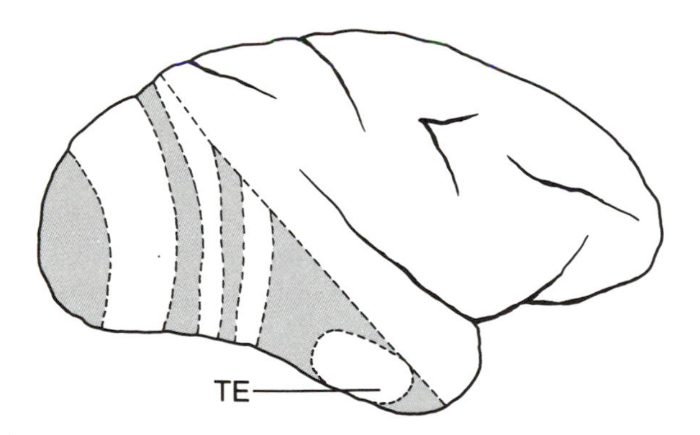

Location of Area TE on the Temporal Lobe of a Monkey Brain

visual areas and for a long time was not considered a visual area. (When this area, called TE—it is on the *te*mporal lobe of the cerebral cortex—was damaged in monkeys, they had great difficulty learning certain kinds of visual tasks, but this is the topic of another chapter.)

The discovery that area TE has prominent visual input is a classic case of serendipity. According to the story, Charles Gross and his associates at Yale University were studying the responses of cells in this area to visual stimuli in the monkey. They used spots of light, moving edges, and bars that are standard elementary visual stimuli. Neurons in area TE responded a little bit to these simple stimuli and not to sounds or touches, so it seemed to be a visual area, but not much of one. After studying one particular cell for a long time with minimal results, they decided to move on to another cell. As a gesture, one experimenter said farewell to the cell by raising his hand in front of the monkey's eye and waving good-bye. To everyone's surprise, the cell responded strongly, firing wildly in response to his moving hand. The experimenters madly cut out different paper shapes of hands and showed them to the cell, which responded most dramatically to an upright shape of a monkey's hand. Cells in visual area TE seem to respond better to complex specific shapes than to simple ones.

There is a very ancient debate about how we see the world. Do we learn to see it as we do, or is it given? The scientific answer seems more and more to be that it is given—determined by the extraordinary architecture of the visual cortex. However, normal visual experience is critically important to the normal growth and development of this architecture; witness the story of Bobby.

Work has really only begun on many of these secondary visual areas, each of which seems to have the basic organization into columns—clusters of cells running from top to bottom that have common functional properties. The great advantage of the columnar structure is that we can organize several dimensions of information in such a small space. Take visual area V1. The two-dimensional surface area of V1 represents the spatial map or layout of the retina—the receptive surface of the eye—and hence the spatial extent of the visual world we see. The large columns running through the cortex contain right-eye or left-eye input. Within each large eye-dominant column are many small columns, each having cells that respond to a different orientation around the compass. It is a four-dimensional array, two dimensions represent-

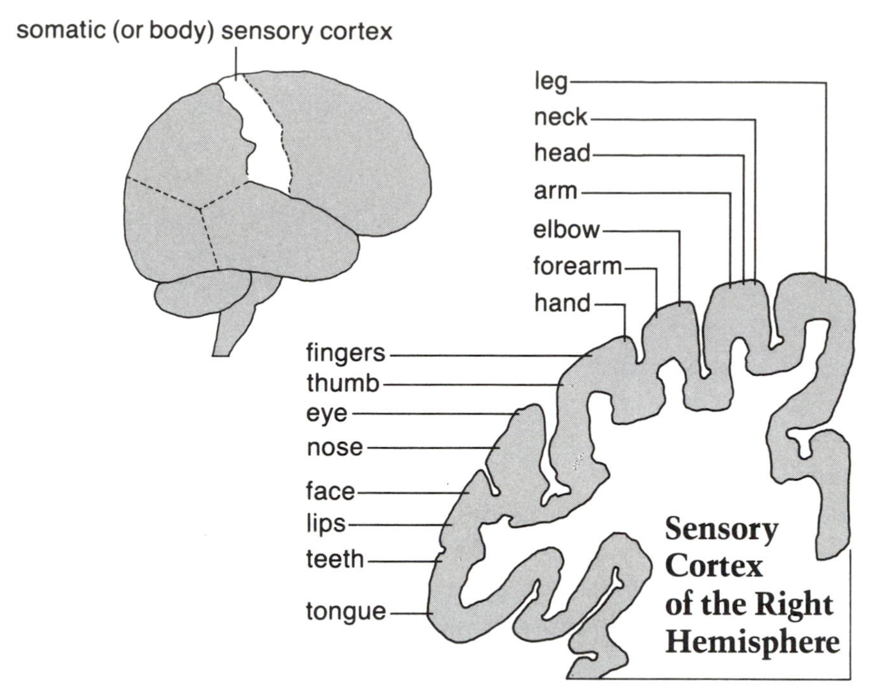

Section Through the Somatic Sensory Cortex, Showing Specialized Areas

ing the entire extent of the area of the retina and two dimensions running at right angles through the cortex to code for which eye is seeing and for the orientation of lines and edges seen.

The area of the cerebral cortex representing skin and body surface, the somatic sensory cortex, seems to be organized in much the same way as the visual cortex, with several separate areas specialized for different aspects of sensory experience from the skin. The nerve cells in a given area respond best to one aspect of touch—light touch and pressure—and in another area respond more to aspects of movement of the fingers and limbs. Like the visual cortex, the somatic cortex is a patchwork quilt; each area is composed of many thousands of little columns of cells that code the many aspects of our sensations from the skin and body.

The fact that the cerebral cortex as a whole is organized into columns was first discovered by Vernon Mountcastle at Johns Hopkins University in his work with cats and monkeys on the somatic sensory cortex. The body surface is laid out along the somatic cortex, and the pattern of representation looks like a homunculus, or little person. The map is complete and highly detailed but very distorted—in humans much of it is devoted to tongue, lips, and fingers, as befits the behavior of *Homo sapiens*. In the rat, which explores its world mostly with its nose and whiskers, the nose and whisker area of the "ratunculus" is much expanded. Each whisker hair has pressure receptors that convey very sensitive information about movements of the whisker to the somatic area of the cerebral cortex in the face region. Each whisker is represented by a separate column of nerve cells through part of the cortex that has the shape of a barrel in layer IV. The sides are composed of a cylinder of cells surrounding a central area with many fewer cells. Immediately outside the barrel there are almost no cells and then the wall of another barrel. When you look at this cortical tissue you can actually see the barrels. Each barrel is a column of neurons coding movement of a single whisker.

The region of the cerebral cortex concerned with hearing, the auditory cortex, is also a patchwork quilt, but we don't yet know what its various areas may be specialized to do. The fact that speech and language involve certain specialized areas of the human cerebral cortex suggests that the auditory areas may be many and complex. What little evidence we have suggests that the auditory areas, like the visual and somatic cortexes, are organized into thousands of tiny functional columns of nerve cells.

The sensory areas of the cerebral cortex—visual, somatic, and auditory—are all organized into columns. But what about the highest areas, the association areas? These are the vast areas of the human cerebral cortex that are neither sensory nor motor (movement) in function but involve more complex cognitive and decision-making processes. Visual area TE, which contained the monkey "hand" cell, is actually thought to be an association area, but too little work has been done on it to know whether it is organized into columns.

There do appear to be complex functional columns of neurons in association areas of the cortex that seem to relate skin sensations and limb movements to vision. Damage to these association areas in people leads to neglect of the relevant part of the visual world. Patients with such damage are reluctant or unable to attend to visual stimuli and to reach out and touch objects. Mountcastle discovered columns of "command" neurons in this region; the cells would become active only when a trained monkey reached out to touch an object that would provide it with a food reward. Moving the animal's arm passively or touching the hand would not fire the cell, nor would the visual stimulus that told the animal it could obtain a reward. It was only when the monkey *decided* to reach out and touch the object to get the reward that the cell fired.

Actually, there appear to be separate columns of cells in this association area for different aspects of the monkey's intentional movement. One column of cells responded when the arm began to reach out, another when the hand manipulated the object, still another when the animal actively moved its eyes to look at the object, and so on. It may be that the "decision" to make voluntary movements originates in this association area. Could this be the cellular basis of free will?

The seemingly ramshackle architecture of the brain becomes clearer as we examine it more closely. The large visual "room" is divided into many smaller ones, the visual areas of the cortex, each seemingly with a special job of its own. The "architect" who designed the cortex was exceedingly fond of columns: each room is filled with columns, each one representing a fine shade of visual experience. But what did the architect use for building blocks?

3

Neurons: The Building Blocks of the Brain

NEURONS, the nerve cells that are the major constituent of the brain, are in many ways the most remarkable cells in all of biology. Most neurons in the brain are very tiny, some no larger than a few millionths of a meter in diameter, but their numbers are legion.

The basic job of the neuron is to process information and convey it to other neurons in the brain and ultimately to generate behavior and experience. It is often thought that nerve cells carry information by sending electrical signals along their nerve fibers to other neurons. Such is not the case, as we will see. They do indeed generate electrical fields, quite large ones that can easily be recorded from the surface of the scalp in humans and other animals. But the nerve impulse is not at all like electricity in a wire—it travels along the nerve fiber much more slowly and is a process that involves the exchange of chemical particles between the outside and the inside of the fiber.

The brain is often likened to a computer and the nerve cells that make it up to the elements in a computer, but the analogy is not very accurate. The brain is living—it can grow and change, but a computer can't—and it is infinitely more complex than present-day computers. Indeed, each nerve cell in the brain functions more like an entire computer all by itself. A neuron is not simply on or off like an element in a computer; it is always processing the information it receives from thousands of other nerve cells and from chemical messengers in the

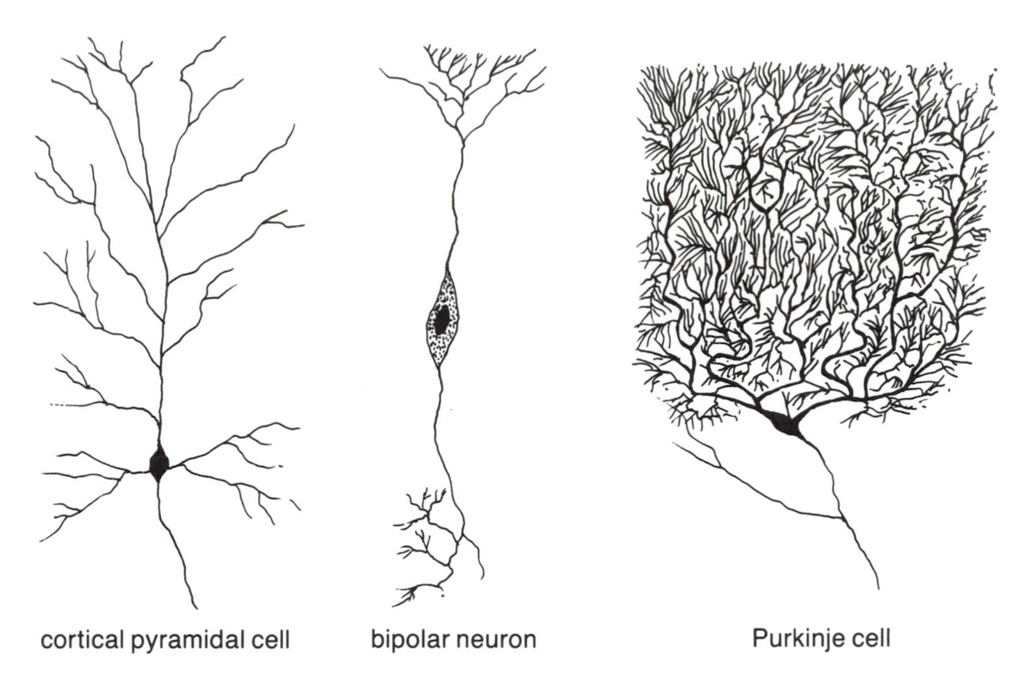

Three of the Many Types of Neurons

bloodstream and is always in communication with many other nerve cells. To appreciate just how extraordinary the nerve cell is, we must look at the nature and evolution of all biological cells and how they came to be, from the first cells to the first nervous systems.

The origin of life on earth is a fascinating puzzle. We can only make educated guesses about it, however, because we have no direct evidence in the form of fossil records of anything more ancient than bacteria. In the beginning, as far as we know, the first cells to form were the bacteria, primitive cells without a nucleus. In bacteria the DNA (short for deoxyribonucleic acid), which is the genetic material, is present as a single, long, convoluted molecule intertwined within the entire cell. In more advanced cells, like those of which we are made, the DNA is contained within a specialized central structure, the nucleus.

The singularity of DNA indicates that all forms of life that exist in the world today — bacteria, plants, animals, you and me — are likely descended from just one original single cell line. How this ancestral cell, destined for an incredible future, began is really still unknown. What follows is our best guess. We suspect that in the beginning the oceans were filled with a large quantity of amino acid molecules—the organic soup of the early world. These molecules are the basic building blocks of life, but they are unable to reproduce themselves.

At some point a remarkable molecule appeared. It may not have been the most complicated molecule, but it had the unique ability to make copies of itself. This has been called the replicator molecule. From modern knowledge of chemistry it is not too difficult to see how such a molecule might work. If it is reasonably large, it will consist of smaller units or building blocks. Suppose each of these blocks is itself a relatively simple compound that exerts a chemical attraction on an identical compound that is floating free in the organic soup. If the replicator molecule is a long chain of such compounds, it will build up an identical chain attached to it. If they separate, each new replicator molecule can build another, and on and on. Because the chemical composition of DNA is virtually the same in all living things, it seems we are all descended not just from one cell but from one replicator molecule.

The earth was formed about four and a half billion years ago. The first life appeared about a billion years later. In a very old rock formation, Gun Flint, in Ontario, there are fossils of very early life forms, primitive bacteria. These bacteria were inefficient; they could not obtain and use energy very well. Perhaps a half a billion years later, certain of these bacteria evolved the ability to extract energy from sunlight and carbon dioxide and give off oxygen, as modern plants do. Then the other major cell type evolved—the eukaryote ("good nucleus"). Eukaryotes, which make up all animals and plants above the level of bacteria and have a well-developed nucleus, contain the DNA as well as several other tiny structures.

The mitochondria (from the Greek words for "bread" and "grain"), shaped like tiny footballs, are some of these little structures. They have one primary function, which is simple and essential for the life of the cell: they manufacture energy. All cellular processes require energy, and this energy derives largely from a form of sugar called glucose into which certain food substances are converted by the digestive system. Mitochondria are where glucose and other food substances are metabolized to form biological energy.

Mitochondria are remarkable little organs because they have a life of their own. When a cell divides, each new cell contains some mito-

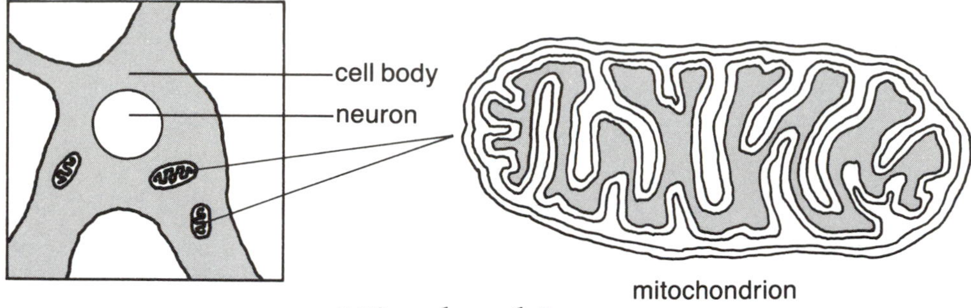

Mitochondria

chondria from the parent cell, which means that they would soon be used up in cell division as a new animal develops. However, the mitochondria themselves divide and form new mitochondria within the cell. Each mitochondrion has its own genetic material—DNA—and reproduces itself. Since making a new mitochondrion is much simpler than making a new animal, its DNA is very much smaller and simpler than the DNA in the cell nucleus. Another astonishing fact about mitochondria is that all your mitochondria come from your mother. The lineage of mitochondria is entirely maternal.

In almost every way, the mitochondrion resembles a bacterium. It has DNA but not a nucleus of its own and can manufacture its own energy. Many biologists now think that the ancestors of mitochondria were free-living bacteria that then entered other cells and became very specialized parasites, or, rather, symbiotes (from the Greek for "living together"). The cell is completely dependent on the mitochondria to manufacture energy for it. The mitochondrion, on the other hand, although it can reproduce, cannot make all the proteins it needs to function properly. Some of these are made by the DNA in the host cell.

The cell membrane separates the cell from the world outside and is the functional boundary of the cell. All commerce that the cell has with the outside world must somehow pass through or act through the membrane. It resembles a soap bubble, being very thin, and consists of fatty acids, which also make up the film on a soap bubble. It seems a rather fragile way to protect the cell from the outside world, but it works.

Various protein molecules are scattered throughout the membrane, literally floating in it. Some may be large enough to extend all the way across the membrane, but most of them tend to be closer to the outside or the inside. A protein molecule that is toward the outside generally stays there; it can float laterally along the membrane but does not move to the inside. The same is true for those that are toward the inside. These protein molecules have chemical side chains that tend to stick out of the membrane, on either the outside or the inside.

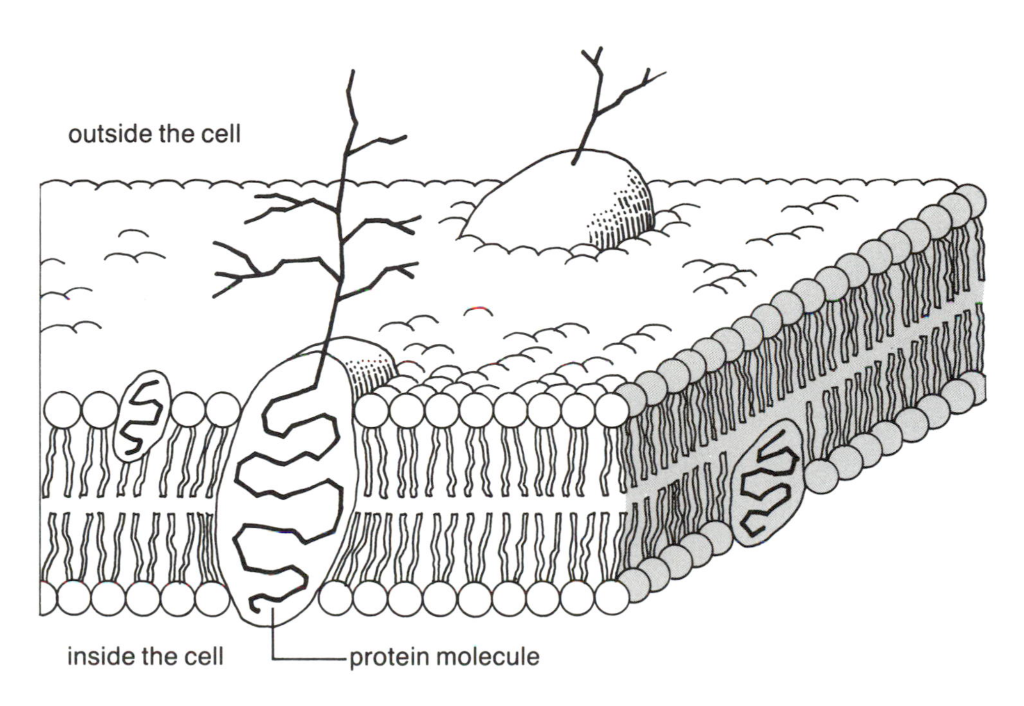

Cross Section of the Cell Membrane

The protein molecules and their side chains are believed to be the chemical receptors on the cell membrane. The notion of the receptor molecule has become fundamental in cell biology and in our understanding of the nervous system. The basic idea is that a given protein molecule recognizes a particular chemical substance. If that substance is present outside the cell, molecules of it will attach to the protein molecules in the cell membrane, much as a key fits in a lock, and cause various changes in the cell membrane and cell.

After the appearance of more sophisticated cells with a nucleus and mitochondria, the pace of life accelerated. During the next half-billion years, cells bonded together to become multicelled organisms. The sponge is a living example of these most primitive multicelled organisms, and it has no nervous system. A sponge does not move—it just sits around. If nutrients are in the water, it lives; if not, it dies.

The first nerve cells developed in animals such as the sea anemone and jellyfish. These organisms represent a great advance over the sponge because they *behave*—the jellyfish, for example, can swim to where nutrients are and capture them.

In order for a multicellular animal like a jellyfish to move, its individual cells must somehow be made to move, and indeed, some cells became specialized as muscle tissue that can contract. But in order for these movements to accomplish anything they must be coordinated, they must somehow be controlled to move together. This task requires nerve cells.

So far as we can determine, nerve cells in every animal from jellyfish to human have the same basic electrochemical mechanisms for conducting information. It appears that the primitive mechanism of the jellyfish nerve cell worked so well that it became fixed in evolution. In order to generate more complicated and adaptive behavior, all you need to do is put more of these nerve cells together in more complicated ways.

Nerve cells are "good nucleus" eukaryotic cells and resemble all other cells in the human body in most ways. Each has a nucleus containing the DNA, a cell membrane covering the entire cell, and mitochondria and other little structures. Nerve cells differ from most other cells in just a few ways. Most important, they have become specialized to conduct information to one another by way of long fibers that extend

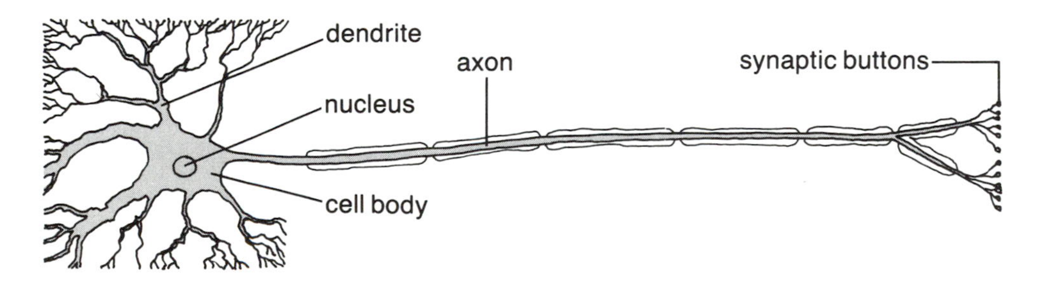

A Typical Nerve Cell

out from the cell body. There is just one fiber—the axon (from the Greek for "axis" or "axle")—that carries information to other cells. All the other fibers that extend from the nerve cell body are dendrites (from the Greek word for "tree") and receive information from the axons of other nerve cells. In humans, the nerve cell axon can be as long as three feet, but the dendrites are always quite short, less than a millimeter.

In the nineteenth century it was generally accepted that all living things are made of cells, from the most primitive single-celled organisms—the bacteria—to us. Indeed, this cell theory of life became the great unifying principle of biology. Surprisingly, even at the beginning of the twentieth century many anatomists doubted that the brain was composed of individual cells and felt it was the one exception.

To see cells in a microscope it is necessary to color or stain them with dyes. If a piece of brain tissue is stained with a dye that colors all parts of all cells, it will look like a continuous mass of tissue, a tangled web of fibers and processes with cell nuclei scattered throughout. In the late nineteenth century an Italian anatomist, Camillo Golgi (1844–1926), developed a special kind of stain that would color only an occasional nerve cell in brain tissue but color it completely. Using this stain, it is possible to see complete nerve cells with all their fibers and processes. Interestingly, the stain was discovered by accident. The story goes that a cleaning woman threw a piece of brain tissue from Golgi's desk into a waste bucket that happened to contain a silver nitrate solution. When Golgi returned and found the tissue, it turned out to be the first successful Golgi stain.

Golgi himself still did not believe that the brain consisted of individual nerve cells. But another anatomist, Santiago Ramón y Cajal (1852–1934), a Spaniard, systematically applied Golgi's method to the brain in animal studies and established that all parts of the brain are, in fact, composed of individual nerve cells. Cajal began the immensely complex task of working out the patterns of interconnections among the neurons in the brain—the wiring diagrams, you might say.

The neuron is the functional unit of the brain. It receives information at its dendrites, processes it in the cell body, and sends it out to other neurons and cells along its axon. The axon separates into a number of small fibers that have terminals. Each of these terminals forms a functional connection to another cell, the synapse. Actually,

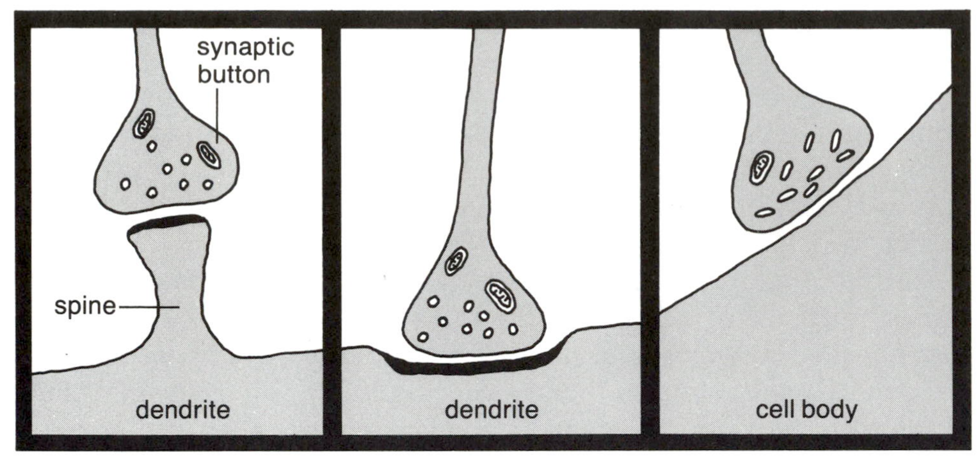

Spine Synapse **Excitatory Synapse** **Inhibitory Synapse**

Typical Synaptic Connections

there is a very tiny space in the synapse between the axon terminal and the cell body or dendrite of the cell on which it terminates. As far as we know, a neuron communicates with other neurons (or muscle or gland cells) only by way of these tiny synaptic connections. The synapse is thus the functional connection that permits one neuron to communicate with another.

A given neuron in the brain may have several thousand synaptic connections with other neurons, and in turn make synaptic connections to many other neurons. If the human brain has 10^{11} (100,000,000,000) neurons, then it has at least 10^{14} (100,000,000,000,000) synapses. However, it is noteworthy that the number of *possible* synaptic connections among the nerve cells in a single human brain is virtually without limit.

One major difference between neurons and other cells is that after a certain point nerve cells never reproduce new nerve cells. As the organism develops from the fertilized egg, nerve cells develop and multiply to form the brain and nervous system (at the amazing rate of 250,000 per minute over the nine months of development of a human fetus). By birth this process has almost ceased, and, in fact, in some parts of the brain there are more cells present at birth than later, because some die out. We still have no clear idea why no nerve cells are formed after birth; the ultimate answer is likely embedded in the genes.

However, we can assume that since particular behaviors and behavior sequences are the result of the development of particular sets of connections among the nerve cells of the brain (specific wiring diagrams, as it were), if the nerve cells were to divide and form new cells, these patterns of connections would be lost.

There seems to be one rather astonishing exception to this rule. The song region in the brain of the male canary grows to double its normal size in the spring as the bird learns its song to attract a female. After the mating season, the song area shrinks and the song is forgotten. Next spring, the song area grows again as the canary learns a new song. Imagine what life would be like for us if our brains were like those of male canaries. Each year we would forget everything we had learned and have to start all over again! The growth of new song circuits each year in the canary brain seems to be a very specialized development to accommodate a very special form of learning (and it *is* learning—the canary learns a different song each year). The remainder of the canary brain does not change. The circuits responsible for other aspects of its behavior do not grow and disappear, only the song circuits.

So far as we know, no such specialized process occurs in the mammalian brain. But although the full complement of nerve cells is present shortly after birth and no new nerve cells are formed, experiments suggest that even in mammals, nerve cells can be made to "sprout" and grow new axon terminals. Perhaps someday they even can be induced to grow new nerve cells.

In common with other cells, the body of the neuron is a chemical factory, making many different substances that the nerve cell will use. Nerve cells have the special problem of getting chemicals from the cell body out along the axon to the synapses. Before the electron microscope was developed, many scientists thought that the interior of the axon was simply a continuous jellylike substance without structure. Now we know that it is filled with tiny tubes that move chemicals from the cell body out the axon to the synaptic terminals. It is easy to show this, because if an axon is tied off, it will gradually swell up on the side of the tie toward the cell body. Chemical substances cannot cross the tie and therefore build up. The synapses also take up chemicals and transport them backward along the axon to the cell body.

The movements of chemicals up and down the axon are relatively slow, typically taking hours to cover the distance from the cell body

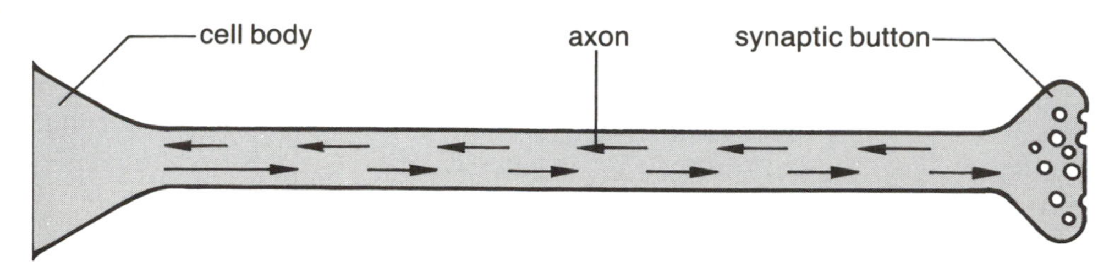

Movement of Chemicals Along an Axon

to the synaptic terminals. These movements are essential for the neuron to function, but such is not the way the nerve cell communicates with other cells. A nerve cell talks to other nerve cells very rapidly, in a few thousandths of a second.

The most important part of the neuron is the synapse; only nerve cells and their target cells have synapses. When you look at a section of brain tissue in a light microscope (a light microscope does not magnify things nearly as much as an electron microscope, which uses a beam of electrons rather than ordinary light), you see a bewildering variety of contacts among nerve cells. But there are basically only two types of synapses — the chemical synapse and the electrical synapse. Most synapses in the mammalian brain are chemical, so with apologies to invertebrates that have some very interesting and specialized electrical synapses, we focus here on the chemical synapse.

All chemical synapses have common features that serve to identify them. The most obvious is the presence of a large number of small spheres, or vesicles, clustered in the presynaptic terminal at the very end of the axon, which is to say just before the synaptic space. The vesicles are believed to contain the chemical neurotransmitter substance for the synapse. The region of the synapse just adjacent to the terminal on a target neuron, the neuron on which the synapse occurs, has a dense band, which looks dark in the electron microscope, along the cell membrane that defines the extent of the synapse. In between the presynaptic and postsynaptic membrane is a space — the synaptic space. This space is always present and uniform; it is a very tiny space, but nonetheless a space.

When a synapse is active and transmits information, the vesicles are believed to release their content of neurotransmitter chemical into the synaptic space. The chemical molecules diffuse across the narrow synaptic space and attach to receptor molecules on the outside surface of the postsynaptic membrane. This results in activation of the target cell.

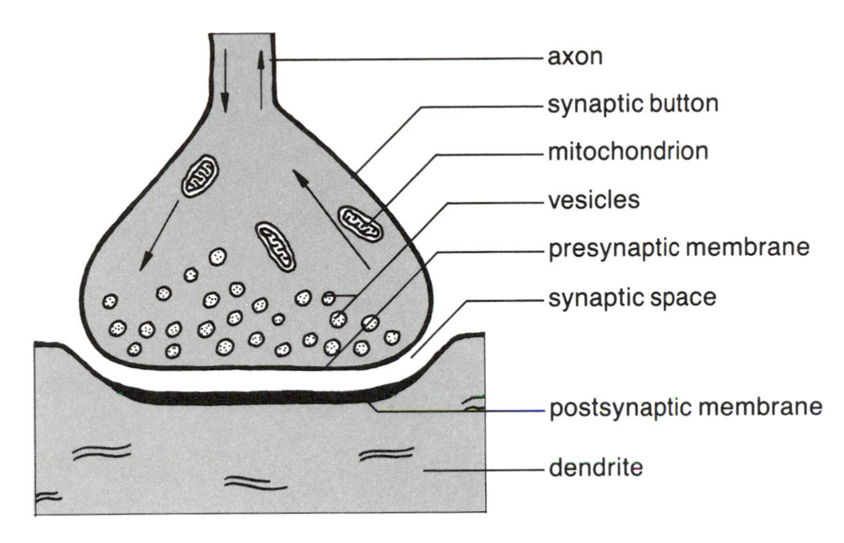

Detail of a Synapse

Pretend that you are tiny enough to stand inside the synaptic space on the postsynaptic target neuron membrane and are looking up at the terminal. When the nerve impulse arrives, you would see large openings developing in the terminal, and a hundred or so of the vesicles will dump out a few thousand molecules of chemical transmitter substance. It would seem like a brief local rainstorm.

Nerve cells use two quite different "languages" to communicate with each other. One of these is the nerve impulse, also called the action potential, a process that develops in the axon at the point where it exits from the nerve cell body and travels out the axon to the axon terminals. When it reaches the axon terminals that form synapses on other neurons, the action potential ceases; it is finished. But it has done its job because when it arrives at a terminal, it triggers an entirely different process: transmission of information across the synapse to the target neuron, the neuron receiving the information. Synaptic transmission is the other language of the neuron. As we have seen, the process is a release of chemical transmitter molecules from vesicles in

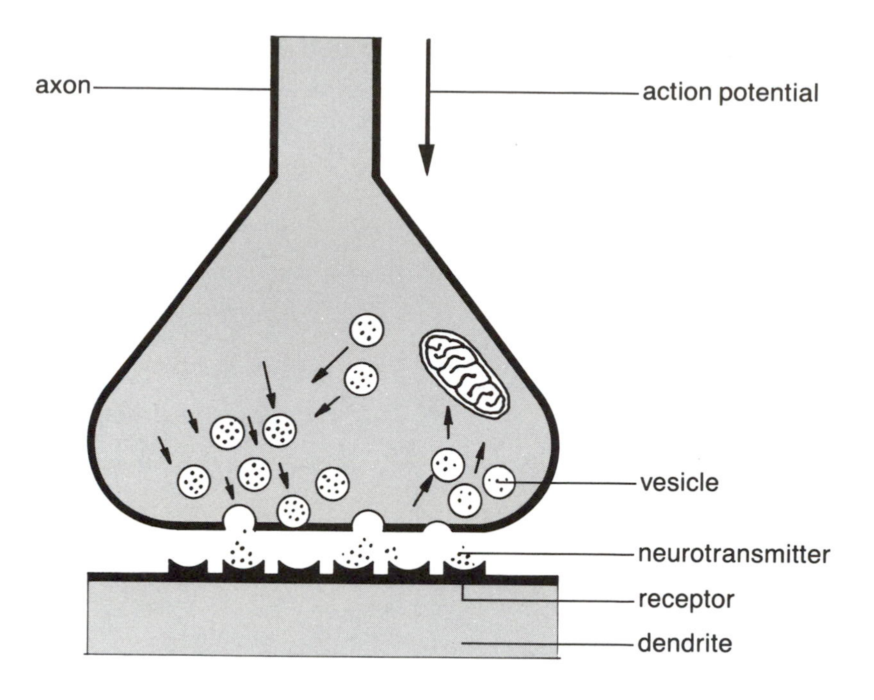

Detail of Synaptic Transmission

the synaptic terminal to attach to the chemical receptor molecules on the target cell membrane.

The nerve impulse is all or nothing; once it begins at the beginning of the axon, it develops full-blown and travels all the way out the axon. It is always the same in all axons except for its speed. How fast it moves down the axon is determined by the size and other properties of the axon—it can range from about one mile per hour to one hundred fifty miles per hour.

Synaptic transmission is entirely different. It is graded rather than all or nothing. The amount of a synaptic action depends on many factors, among them the number of chemical molecules released and the number of synapses on a given neuron that are active at any particular moment.

Most modern computers are digital computers, in which each element is either on or off, just as the action potential in the nerve cell occurs or does not occur. However, there are also special-purpose analog computers that use varying or graded amounts of electricity, much like the synapses on a nerve cell. Each nerve cell is like a computer that is both digital and analog.

This, then, is the nerve cell—the building block of the brain. All nervous systems are made up of these same basic neurons, from the few thousand in simple invertebrates to the many billions in our own brains. The kind of creature we are, whether a simple reflex machine like the sea anemone or a being with the truly awesome power of the human mind, is determined by the number of neurons and how they are put together.

We focus now on the nerve impulse. The special properties of the nerve cell that allow it to conduct information to other cells reside in the membrane that covers it. The general structure of the nerve cell membrane is the same as other cell membranes, and it shares the general function of all cell membranes, protection and exchange of chemicals between the inside and the outside. One other key property of the nerve cell membrane is common to other cell membranes as well: tiny holes or channels through the membrane that allow certain small molecules to pass through. At the time the first nerve cells developed in animals like the jellyfish, certain of these channels became specialized in such a way that a message—a nerve impulse—could be conducted out the axon of the nerve cell to influence other cells.

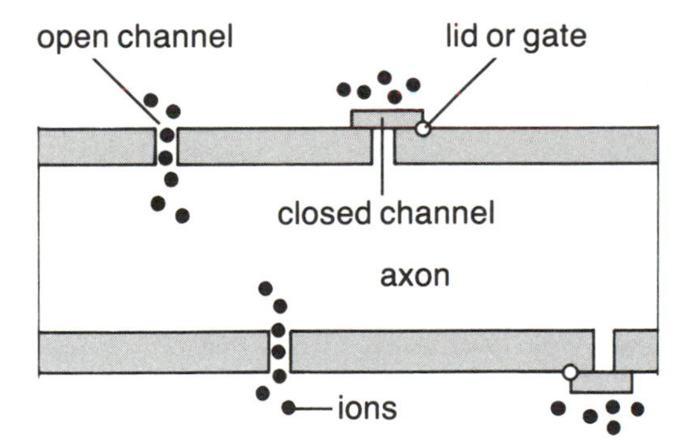

Diagrammatic Section of Axon Membrane

The nerve impulse is a movement of chemical particles through the axon membrane, and this occurs only at the little region of the axon where the nerve impulse is at a given point in time. The major

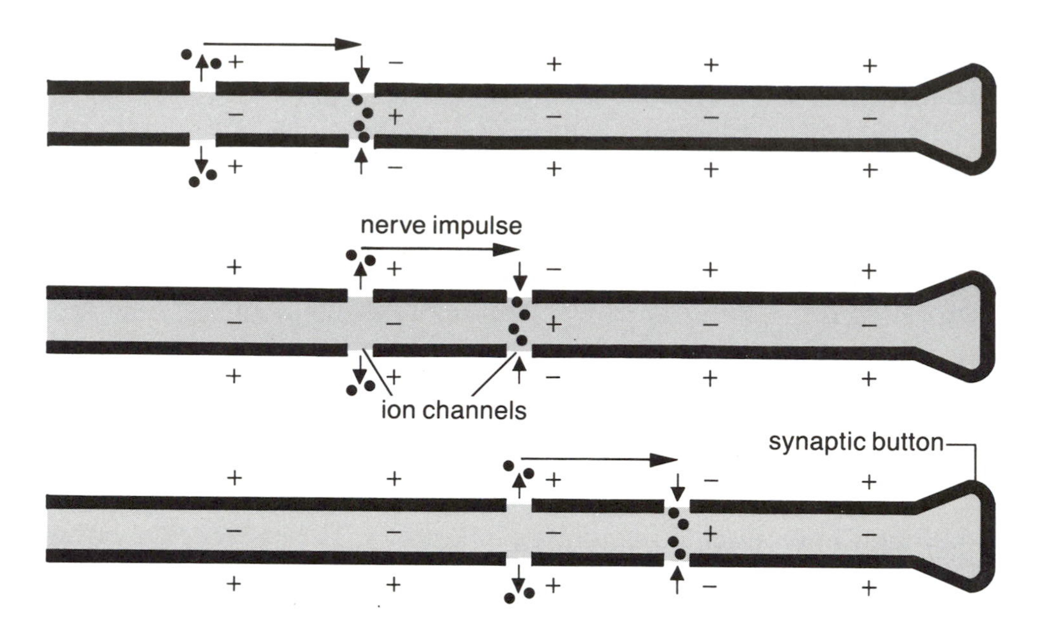

Movement of Nerve Impulse Down an Axon

chemical particle that crosses the axon membrane in the nerve impulse is the sodium atom, but in the form of a charged atom, an ion.

Think of the axon as a long, thin tube covered by the nerve cell membrane. Inside the membrane is the interior substance of the axon. Outside the membrane is the extracellular fluid — tissue fluid. The interior substance and the fluid outside have quite different chemical make-ups. The substance inside the axon has many protein molecules, for example, and very little sodium. The outside fluid has little protein but a considerable amount of sodium. The fundamental event in the nerve impulse is the movement of sodium molecules across the cell membrane from the outside to the inside. They move through the little channels in the membrane. These sodium channels are normally closed, but when the nerve impulse develops, they pop open briefly to allow sodium to move in. To understand how they do this, we must turn to the electrical aspects of the nerve impulse. Nerve cells generate electric currents, so much so that it is easy to record the electrical activity of the brain by putting wires on the surface of the scalp. To understand the electrical aspect of the nerve impulse, we must become chemists for a bit. The sodium particles that move from the outside to the inside of the axon membrane during the action potential are in the form of particles called ions, which have an electrical charge.

Table salt is composed of sodium and chloride ions. When you dissolve salt in water, the sodium and chlorine atoms are present with their electrical charges, sodium atoms each having a charge +1 and chloride atoms each having a charge −1. You may recall from high school science courses that atoms are composed of a nucleus that has positively charged protons (and neutrons that have no charge) surrounded by shells of negatively charged electrons. In an element made up of the same kind of atoms, like sodium, there are the same number of protons in the nucleus as electrons surrounding it in each atom. It is electrically neutral. The element sodium is a highly toxic and explosive silvery metal. It is especially reactive with water: toss a piece of sodium into water and it will burst into flames and explode.

The element chlorine is a poisonous greenish gas. Its atoms have the same number of protons in the nucleus as electrons around it, and it is an electrically neutral but highly reactive element. Why are these simple elements so volatile?

The electrons around any atom exist in shells. The innermost shell requires two electrons to be complete and the next several shells each require eight. Atoms are most stable if the outer shell of electrons has its full complement. Helium, the inert nonexplosive gas used for dirigibles, has just the complete inner shell of two electrons (balanced by two protons in the nucleus). The most reactive atoms are those that have only one electron in the outer shell (which they try very hard to give up) or are missing only one in the outer shell (which they try to grab from other atoms). Sodium has only one in its outer shell and chlorine has seven in its outer shell. If sodium and chlorine elements are combined, they react immediately and violently together. Each sodium atom gives up its one outer electron to a chlorine atom, which takes up the electron to complete its outer shell of eight. The result is table salt! Each sodium atom now has one less electron around the nucleus than it has protons in the nucleus and has an electrical charge of +1. Each chlorine atom now has one *more* electron than it has protons and has an electrical charge of −1. They are now ions—atoms with an electrical charge. There is even a change in the name for chlorine when it is in the ionic form, chloride. Table salt is sodium chlori*de,* not sodium chlorine.

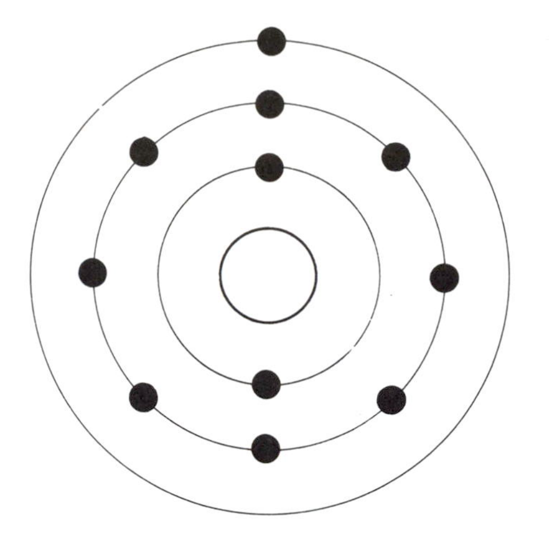

Sodium Atom

11 protons (+)
11 electrons (−)
(electrically neutral)

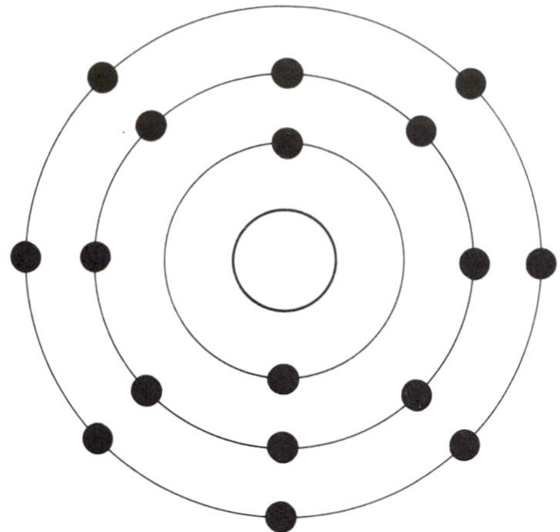

Chlorine Atom

17 protons (+)
17 electrons (−)
(electrically neutral)

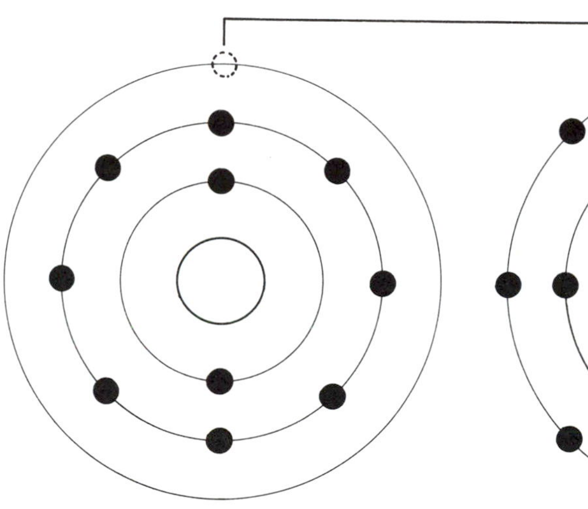

Sodium Ion

11 protons (+)
10 electrons (−)
(lost one electron)

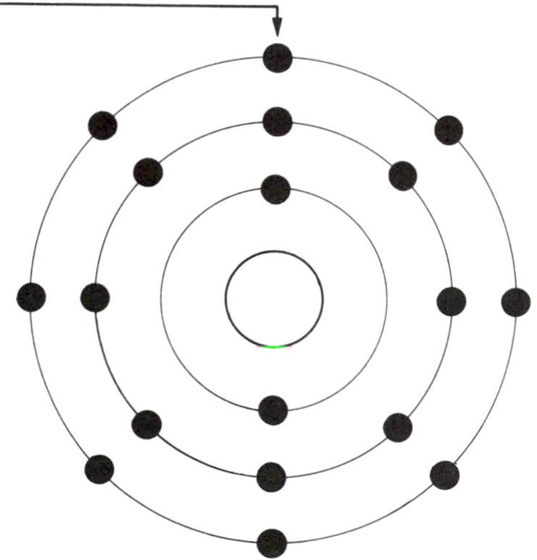

Chloride Ion

17 protons (+)
18 electrons (−)
(gained one electron)

One electron from the outer shell of the sodium atom was lost to the outer shell of the chlorine atom, making them both electrically unbalanced and, therefore, ions.

Chemical substances dissolved in water are typically in the form of ions, having electrical charges. Only a few small ions — sodium, chloride, potassium, and calcium — are involved in the activity of neurons. Protein molecules, the large molecules made up of amino acids, which are the stuff of life, are found mostly inside cells in the body, including nerve cells, rather than outside cells in blood or other body fluids. Dissolved in water, proteins are usually in ionic form and have a net negative charge, like chloride ions. Consequently, there are many more negative charges inside nerve cells and other cells than outside. As a result, there is a voltage difference across the membrane of a nerve cell — the inside is negative relative to the outside. This voltage difference is surprisingly large, nearly a tenth of a volt. If we could hook up a few nerve cells together in the right way they would have as much voltage as a flashlight battery, even though the nerve cells are very tiny. This is not merely an academic point: the electrical eel can generate several hundred volts, enough to kill its prey fish, because it has specialized nerve cells hooked up together in just this way. It is not entirely beyond the realm of possibility that the mechanism the nerve cell uses to generate electricity might someday become a useful source of power.

The nerve impulse, then, is a process that involves primarily sodium ions, which have a positive electric charge. When a nerve cell is at rest, almost all the sodium ions are outside the cell. But because of the protein molecules, the inside of the cell is electrically negative relative to the outside, so there is a very strong electrical force trying to pull sodium ions into the cell. In the nerve cell membrane are sodium channels, holes that allow sodium ions to pass through.

At rest, these sodium channels have gates that are closed so the sodium ions cannot rush into the cell. But when the nerve impulse develops at a particular place on the axon, the sodium channel gates in that region pop open very briefly and sodium ions rush in. The voltage just inside the membrane at that place shifts from negative to positive because the sodium ions have positive charges that they carry inside. This is the nerve impulse — a brief local inrush of positively charged sodium ions.

When the nerve impulse occurs at one place on the axon membrane (and it occurs all the way around the axon membrane at that place), the closed gates on sodium channels, which are voltage-controlled just like electrical switches, pop open briefly and the nerve impulse moves to this next place, and so on all along the axon. This is how the nerve impulse travels along the axon.

Why do the closed gates on the sodium channels suddenly pop open as the nerve impulse develops? When the voltage across the membrane changes just a little so that the inside becomes just a bit more positive than its normal resting negative value, this slight positive shift triggers the switches on the closed sodium gates and pops them open. The sodium channel gates are thus electrically controlled and have a threshold or trigger level that is just a bit less negative voltage than the value of the resting membrane potential on the inner surface of the membrane. When the inner surface of the membrane reaches this threshold voltage level, the sodium switches are activated and pop open in an all-or-none fashion, generating the action potential. Consider the closed sodium channels just next to where the nerve impulse is occurring on the axon membrane. Positively charged sodium ions are rushing in at the place where the nerve impulse is occurring. They build up inside the axon membrane under the closed gates just adjacent, which makes the potential just a little bit more positive inside the membrane there. This triggers the closed sodium gates to open there.

Under normal conditions the action potential starts at the beginning of the axon where it leaves the cell body and travels out the axon to its synaptic terminals on other neurons. But how does it start? Remember that the neuron itself has many dendrites that are receiving showers of chemicals across the synaptic spaces that connect to them from the axons of other neurons, and all these synaptic connections cause small changes in the electrical potential inside the cell membrane. But the cell body and dendrites typically do not have voltage-controlled sodium channels, except on the axon. So when the membrane potential of the cell body becomes enough more positive inside than at rest, the closest voltage-controlled sodium channels are those at the beginning of the axon, and they consequently pop open first and the nerve impulse develops there to begin its journey down the axon.

A remarkable fact about the nerve impulse is that it requires no biological energy to operate—it is close to being a perpetual motion machine. But there is an eventual price. Each time the nerve impulse occurs, sodium ions move into the axon. What happens to them? They are inside and cannot move out. There is another mechanism in the cell membrane called an ion pump, which actually pumps the sodium ion back across the cell membrane to the outside. It operates much more slowly than the nerve impulse and is always at work pumping sodium ions out. Because the electrical force is always trying to pull sodium ions in, the pump has to work against this and use a considerable amount of biological energy. If the energy system in the nerve cell (in the mitochondria) is poisoned, which stops the pump, nerve impulses will continue to occur for hours, but as sodium accumulates inside, the cell will eventually run down and cease to function.

Although the basic mechanism of operation of the nerve impulse involves movements of ions across the cell membrane, and changes in the electrical potential of the cell membrane are critically involved, the nerve impulse itself is not an electric current. Because the cell membranes of neurons do have changes in their electrical potentials, however, they generate electrical fields that can be quite large and easily recorded.

So the action potential travels out the axon to its synaptic terminals and ceases to exist. But when it reaches the terminal, it triggers the quite different process of synaptic transmission: vesicles in the terminal release chemical transmitter molecules, which diffuse across the synaptic space and attach to the receptor molecules on the outside of the target cell membrane on the other side of the synapse. Depending on the type of chemical transmitter and receptor molecules, the synaptic action on the neuron will excite or inhibit the neuron—increasing or decreasing its activity. When the transmitter molecules attach to the receptor molecules, the activated receptor molecules cause changes in the cell membrane that in turn result in excitation or inhibition.

Synaptic actions on a nerve cell from other nerve cells excite the neuron by causing its membrane potential to become a little more positive inside than it is at rest. They are graded and can range from no detectable action to a large action. If enough of these synaptic excitations occur to reach threshold, the sodium gates at the beginning of the axon are triggered to open and the nerve impulse develops. The neuron "decides" to generate an all-or-nothing nerve impulse and the whole process is repeated. If not enough synaptic excitation occurs, the neuron "decides" not to generate an action potential.

If you wished to prevent a neuron from generating nerve impulses—firing, as it is often termed—what would be the simplest way to do so? Make the voltage inside the cell membrane even more negative than its normal resting level. This is just how synaptic inhibition occurs. When inhibitory synapses on a nerve cell are active, they make the cell membrane potential more negative than normal and it cannot reach the nerve impulse threshold. A common way inhibitory synapses do this is by briefly opening some closed chloride channels in the cell membrane. Chloride ions are mostly outside and have negative charges. When they move in across the membrane, they make the membrane even more negative inside than it is at rest.

These two processes, synaptic excitation (making the interior of the membrane a little more positive than at rest) and synaptic inhibition (making the interior of the membrane a little more negative than at rest), are the basic synaptic actions on nerve cells. A given neuron is continuously bombarded by hundreds to thousands of these synaptic actions exerted by other nerve cells and has a fluctuating level of excitation or inhibition. When the excitation is enough to trigger the opening of sodium channels at the beginning of the axon, the nerve impulse develops and travels down the axon to its terminals, where it produces synaptic actions on other neurons.

The fundamental interactions among neurons occur at synapses, the functional connections that neurons make with one another. You may recall our discouraging statement that no new nerve cells develop in the human brain after birth. That's the bad news. The good news is that new synapses do seem to grow and develop. These connections between neurons are what form the circuits and networks in the brain. The major circuitry is established by birth, but the details and fine tuning of the circuits continue to develop throughout life. Indeed, experience itself can cause new synapses to grow. Experiences, then, can shape the brain.

If you were to look at the general structure of the nervous system in such diverse animals as the earthworm, the ant, the octopus, and the human, you would find that although nature has experimented endlessly with different kinds of nervous systems, the basic mechanisms of action of the nerve cells in all these creatures are the same. What is wildly different is the patterns of interconnections among the nerve cells.

Many invertebrates, such as the clam or the lobster, have relatively simple nervous systems with only a few thousand nerve cells. In these simpler animals, many of the neurons can be individually identified by their location and appearance, and many can be shown to have particular functions. The vertebrate nervous system has taken a different tack. Each region of the brain, even in a mouse, has thousands to millions of neurons. These can be categorized into just a few types on the basis of their appearance, but the number of each type of neuron is vast.

The enormous increase in the number of neurons in the vertebrate brain is the ultimate reason for the development of more complex aspects of behavior and experience—the extraordinary increase in intelligence that occurs throughout the evolution of the vertebrates. The mind cannot exist in a single identified nerve cell, or even in many thousands of them; it is the product of the interaction among the myriad neurons in the vertebrate brain.

4

The Chemical Brain: The Molecule Is the Messenger

OUR VIEW OF THE BRAIN keeps getting smaller, from the different "rooms" of the brain, to the columns, to the neurons that are the building blocks. But to reach the key level in the brain's operation, we need to go down one level further—to the chemical molecules that make these nerve cells work.

Shape or architectural structure is the basic principle of operation of the brain at the level of the chemical molecule. Neurons talk to each other by releasing certain chemical molecules, the chemical messengers, or neurotransmitters, at the synapses. In the last chapter we left a tiny observer standing in the synaptic space of an active synapse under a shower of chemical molecules. When these molecules reach the target-cell membrane, the postsynaptic membrane, they attach to chemical receptor molecules that are a part of the postsynaptic membrane. These are large protein molecules that stick out from the membrane. They have particular shapes. The chemical transmitter molecules raining down on the postsynaptic membrane also have particular shapes, shapes that fit into the receptor molecules, much as a key fits into a lock.

The chemical transmitter molecule is the ultimate unit of action in the nervous system. Earlier we thought that there were only a very few such chemicals, perhaps three or four. It now appears that there may be hundreds of different chemical messenger molecules. A given type of molecule is released at a particular synapse, but there are a number of different chemicals at different synapses. Even more nu-

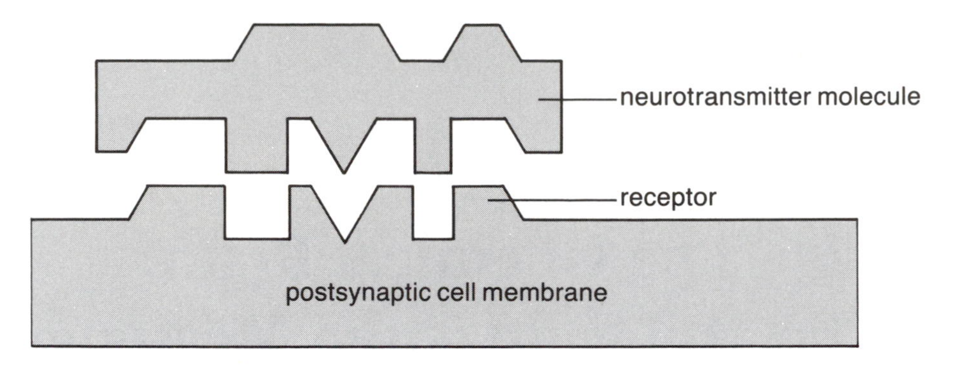

Schematic Diagram of "Lock-and-Key" Mechanism

merous are the chemical messengers released into the bloodstream by glands—the hormones. Hormones also act on chemical receptor molecules in nerve cells and other cells. But they don't act just at synapses, they act on receptor molecules wherever they might be.

For example, oxytocin is a hormone released by the pituitary gland into the bloodstream. When it is time for a baby to be born, it is released and causes the muscles of the uterus to contract, signaling the onset of labor. The muscle cells of the uterus have receptor molecules that recognize oxytocin and only oxytocin. Other muscles in the body are not acted on by oxytocin because they don't have any receptor molecules for oxytocin.

There is only one more step in the working of the chemical synapse. When the messenger molecules attach to their receptor molecules on the target neuron, they trigger the receptor molecules into action. Some receptor molecules are associated with ion channels. One type opens the lids on sodium channels and causes the target cell to become more excited, sometimes to the point where it will initiate a nerve impulse in its axon. Another type of receptor is associated with a different ion channel, say a chloride channel. When it is triggered by its chemical messenger, it opens lids on chloride channels. This makes the cell less excited—in other words, it inhibits the activity of the cell. Chloride ions are negative, and when they move into the cell through their channels they make the inside of the cell membrane even more negative than normal, as we saw in chapter 3.

Still other receptor molecules act more indirectly through what are called second-messenger systems to change the activity of the cell. (The first messengers are the chemical transmitters released at synapses.) When these receptors are triggered by their chemical "first messengers" they activate the second messengers, which act inside the cell and may cause the cell to manufacture more chemicals or more hormones or even make more receptor molecules. The second-messenger system can actually act on the genetic material, the DNA, of the neuron to cause long-lasting or perhaps permanent changes.

The kinds of effects receptor molecules can have on their cells are legion. But they only exert these effects when their particular chemical first-messenger molecules join with them or when drugs that mimic the messenger molecules combine with them. As with the saying "If the shoe fits, wear it," if the molecule fits, the receptor will attach it and be triggered into action. This is the primary reason why very tiny amounts of many drugs have such powerful effects on the brain and mind—they have shapes that resemble the shapes of normal synaptic transmitter chemicals and fool the receptors into believing they are the normal message molecules.

One of the most extraordinary stories about the shape of molecules and the chemical brain concerns a drug that has a very long history in human affairs: opium. This extract of the poppy plant has been used for thousands of years to relieve pain and to induce feelings of intense pleasure. Pain and pleasure are among the most compelling forces that drive human behavior. They are biological imperatives that can override all other factors. Small wonder a drug that can both block pain and produce pleasure has played a major role in history, from ancient times to the opium wars of China to the heroin addict of today.

The major active ingredient of opium is morphine, purified in the early nineteenth century and later synthesized in the laboratory. Morphine is a relatively simple molecule consisting of several atoms having a particular shape. Patent medicines of the late nineteenth century were loaded with morphine, and by the beginning of the twentieth century one out of every four hundred Americans was addicted, which led to the passing of the Narcotic Act in 1914.

The study of morphine actions on the brain led to the discovery of an entire new class of messenger molecules in the brain that act both as synaptic transmitter chemicals and as hormones: the endorphins. Morphine is perhaps the best understood of all drugs that act on the brain. Part of the reason for this is that similar chemical molecules can be made in the laboratory that have very specific antagonistic actions to morphine. Naloxone is the most potent of these—at very low doses it rapidly and completely reverses the effects of morphine. A heroin addict who is about to die from respiratory failure because of an overdose of heroin will be fully awake and recovered in minutes after an injection of naloxone. The addict will also then immediately exhibit severe withdrawal symptoms. Indeed, police have made use of this aspect of naloxone's action. An addict under the influence of an ordinary dose of heroin may appear to be quite normal, but minutes after a naloxone injection he or she will be in severe withdrawal.

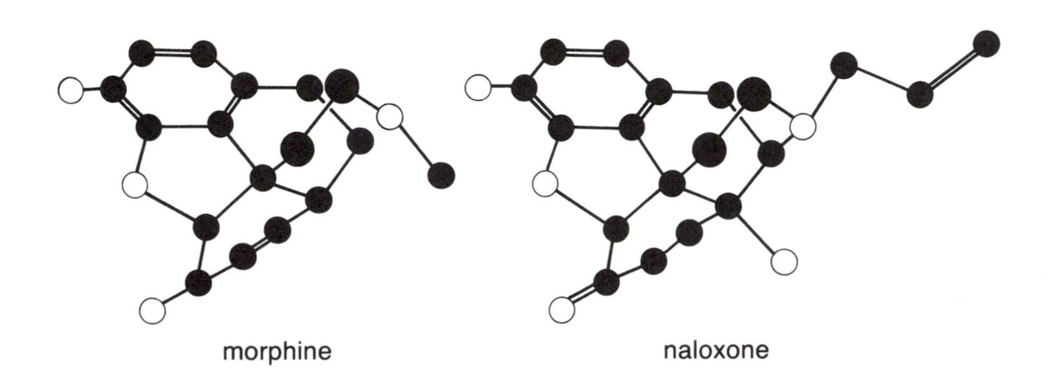

Molecular Models Showing the Great Similarity Between Morphine and Naloxone

Morphine-related drugs and their antagonists are chemically and structurally quite similar. That they have very powerful actions on the brain at very low doses led the pharmacologist Avram Goldstein at Stanford University to propose some years ago that there must be an opiate receptor system on neurons in the brain. Naloxone provides the strongest argument. In terms of chemical structure it is very similar to the morphine molecule. However, it has no detectable effects at all if injected into a normal person who is not addicted to morphine. Indeed, it is so safe that its use has become routine in many hospital emergency rooms. If a person is brought in unconscious with no obvious injuries, naloxone is injected. If the patient is unconscious because of an overdose of heroin, he or she will immediately recover. If the patient is unconscious for some other reason, the naloxone injection will not be harmful.

The opiate receptors, the receptor molecules in the brain that "fit" the morphine molecules, were found in 1974 by Solomon Snyder and Candace Pert, working at Johns Hopkins University. They used some special procedures to enhance the actions of naloxone and showed that radiolabeled naloxone, naloxone made radioactive so it could later be identified with a radiation counter, attached or bound very specifically to receptors on neurons in several regions of the brain.

Why on earth should there be a receptor in the brain that is acted on by an extract of the poppy plant? This receptor system is present in all vertebrate brains, not just the human brain. It is a very old system that apparently developed at about the time the first primitive vertebrates evolved in the ancient seas, long before there was a poppy plant. The obvious answer seemed to be that the body and brain must make its own "opiates" and the receptors are there to be acted on by these normal brain opiates.

However, there are no natural biological substances in the body or brain that are chemically similar to the morphine compounds. There must be other brain chemicals that have an *architectural* similarity—some part of the molecule of a naturally occurring substance must have the same shape as the opiate drugs, a molecular shape that fits into the receptor.

The search was on. Several groups of scientists around the world worked intensely to find these natural opiates. John Hughes and Hans Kosterlitz at the University of Aberdeen in Scotland were the first to succeed. In 1975 they isolated a substance from the brains of pigs that had the same action as morphine. Their procedures were complicated and required the brains of more than two thousand pigs donated from the slaughterhouses of Scotland to yield a small amount of the substance. They named it enkephalin, meaning "in the head." Actually, there are two enkephalins that are closely related chemically and are relatively simple substances.

At first glance these brain "morphines" do not seem to resemble the morphine molecule at all. The enkephalins are peptides—small chains of amino acids. Each is made up of a string of five amino acids. The natural protein substances in meats and other foods are like peptides, only hundreds of times larger—made up of long chains of amino

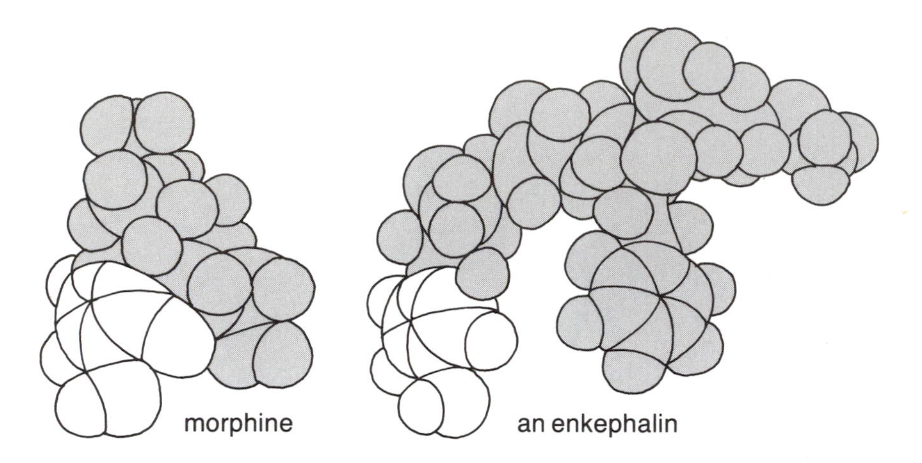

A Spatial Model Showing the Similarity Between Parts of Two Different Molecules

acids. A peptide is a small piece of such a chain. The enkephalins, then, are small peptide molecules. The morphine molecule is not a peptide at all and has a very different chemical make-up.

It is here that chemical architecture provides the answer. In its three-dimensional shape, one end of the morphine molecule closely resembles one end of the enkephalin molecule. The "opiate" receptors in the brain are, of course, not opiate receptors at all; they are enkephalin receptors, acted on by the naturally occurring brain opioids. It just happens that morphine and the closely similar synthetic drugs have a shape that fits the opiate receptor.

Indeed, naloxone, the drug that antagonizes morphine, fits the opiate receptor in the brain even better than morphine. This is why it antagonizes the actions of morphine so effectively. It literally knocks the morphine molecule off the receptor and attaches to the receptor. However, its shape is such that although it attaches to the receptor it does not activate the receptor. Naloxone simply attaches and prevents morphine and other opiates from acting on the receptor.

Morphine, on the other hand, activates the opiate receptors just as do the natural brain opioids to block pain and induce pleasure. Since Hughes and Kosterlitz first discovered enkephalins in 1975, a number of other naturally occurring brain opioids have been discovered. These endorphins (*endo*genous mor*phines*) are all peptide compounds and some are considerably more potent than morphine. The actions of the brain opioids seem identical to the actions of morphine—they relieve pain and induce feelings of pleasure. One might think that these substances would prove to be the ideal painkiller. After all, they are naturally occurring substances in our bodies. Unfortunately, these brain opioids are just as addicting as morphine and heroin. It seems that any substances that act on the opiate receptors in the brain to relieve pain and induce pleasure are addicting.

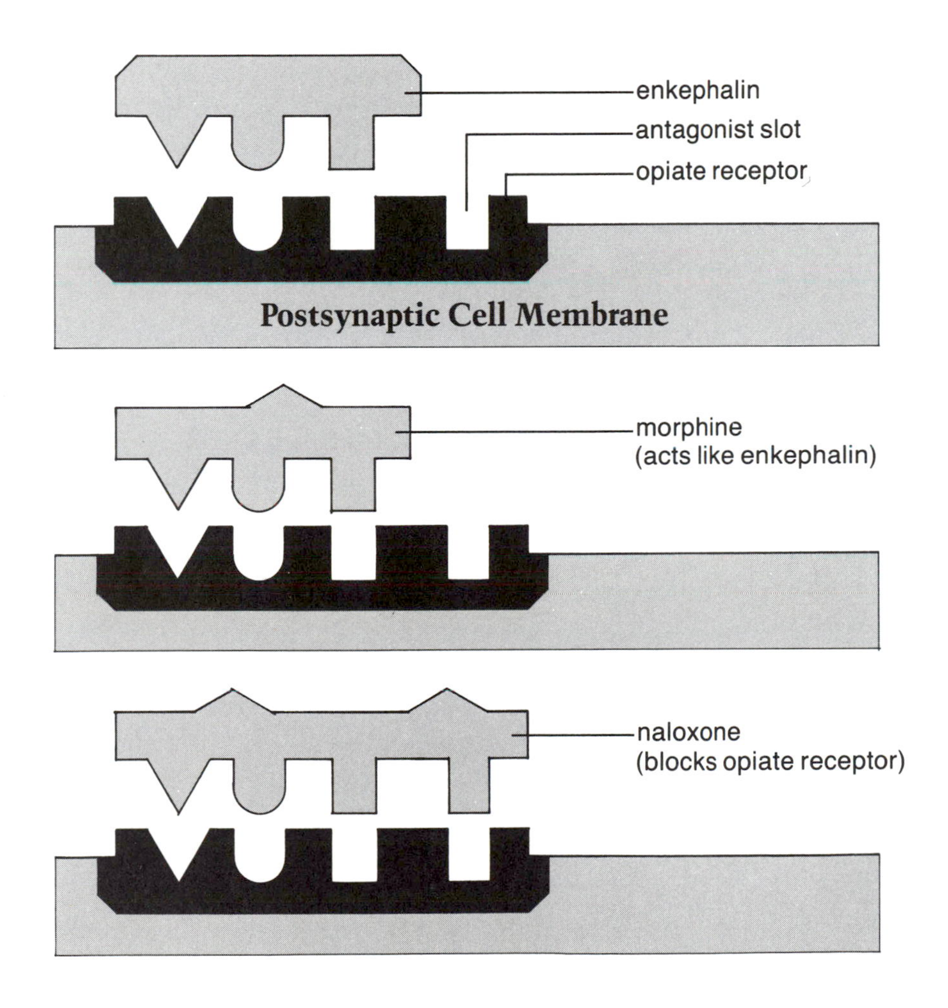

Schematic Diagram Illustrating the Relationship Between Three Different Molecules and an Opiate Receptor

Why do we and other vertebrates have opioid substances in our brains? Animal studies have shown that stress causes these substances to be released into the bloodstream, mostly from the master endocrine gland at the base of the brain, the pituitary. They are released to help counter the pain and suffering induced by stress. In an emergency situation we don't even notice minor injuries that would otherwise be painful.

Naloxone not only antagonizes morphine, it also antagonizes the naturally occurring brain opioids. Injection of naloxone causes clear increases in pain in both animals and humans. Normal people in nonstressful situations given naloxone report no particular subjective feelings, but experiments in which people have to come back for daily doses of naloxone often run into problems because the subjects don't show up after the first dose. They don't report any unpleasant feeling, but their behavior seems to indicate they don't like it.

Avram Goldstein used naloxone to study the possible pleasurable effects of brain opioids. About half of normal adults report that listening to their favorite music gives them a feeling of chills running up and down the spine — the thrill experience. Using Stanford medical students, he injected some with naloxone and others with a control substance. Naloxone caused a significant reduction in the thrill experience.

We might speculate that certain kinds of stressful activities such as long-distance jogging and skydiving are addictive because they induce the release of brain opioids. A common example is the "high" experienced by most serious joggers. They report that if they jog regularly for long distances they develop a strong feeling of joy or euphoria. Interestingly, in childbirth, as labor approaches the level of endorphins in the mother's blood increases dramatically, so that during the birth process both mother and infant have as high as ten times the normal level of endorphins, a dramatic example of the mother's body protecting both her and the infant as much as it can against the pain and stress they are experiencing.

The opiatelike substances released by the body and brain are part of the normal machinery of the brain, as are many other chemical messenger substances. When something goes wrong with one of these chemical messenger systems, it can have drastic effects on the brain and mind. A striking example is schizophrenia — the most devastating form of mental illness. A current view of schizophrenia is that one of the chemical synaptic transmitter systems in the brain, dopamine, has somehow gone seriously awry. The case history of "Judy" is typical of the onset of schizophrenia:

> Judy, a twenty-three-year-old college graduate, was brought to the hospital by her parents because she had developed a severe mental illness. About six months earlier she had quit her first job as a proofreader and became withdrawn and uncommunicative. This in itself was astonishing because Judy had never before been unusually shy or anxious.
>
> When Judy left her job, she knew that something was happening to her, but she didn't know quite what. She became more and more preoccupied with her thoughts. She ruminated on the meaning of existence and religious matters. Her personal appearance deteriorated. She stopped taking care of her hair, using cosmetics, and keeping her clothes clean. Then, a few weeks before she was brought to the hospital, she became convinced that her mission was to save the world from cataclysmic destruction. According to Judy, other people did not know about the imminent threat and did not appreciate the necessity of taking action. Judy said she knew because the information had been directly implanted in her mind by a supernatural power.
>
> She remembered the exact moment when she realized that she had to assume her mission. One morning when she had awakened early, she was standing in her room looking out the window at the early light of dawn. An unusually bright planet was still visible near the eastern horizon. As she watched, the top edge of the sun broke over the horizon and she saw a ray of orange light reach from the sun to the planet. The planet disappeared and the steeple clock began to toll six. She knew then that she had been chosen.
>
> Her enemies knew of her mission because they could read her mind. She tried to occupy herself with trivia to prevent their clairvoyant espionage. They had placed writhing, coiling snakes in her abdomen to stop her. She frequently heard their voices talking about her, swearing at her and plotting how they could thwart her secret plans. Sometimes Judy talked back to her enemies. Many times a day, she received new proof of her role in the great cosmic struggle. She knew that certain events, which others thought were meaningless, were really signs. For example, just before she entered the hospital, a fly had landed on the television and started cleaning its wings while Barbara Walters

was reporting on the satellite pictures from Jupiter. Judy knew then that not much time was left. [From Marvin E. Lickey and Barbara Gordon, *Drugs for Mental Illness: A Revolution in Psychiatry.* San Francisco: W. H. Freeman, 1983, pp. 52–53.]

Schizophrenia afflicts about one percent of the world's population and is independent of race, culture, and life experiences. The fundamental characteristic of severe schizophrenia is disordered thought. It is usually accompanied by a range of false beliefs, or delusions, and auditory hallucinations—hearing sounds that don't exist, particularly voices speaking to the patient.

Schizophrenia has genetic roots and runs in families. If you have an identical twin who is schizophrenic, you have a fifty-fifty chance of becoming so; if you have a schizophrenic brother or sister, you have a one-in-eight chance. If none of your close relatives is schizophrenic, you have the base-rate chance of one in one hundred.

It would be satisfying if we could say that basic research in neuroscience led to a clear understanding of schizophrenia and this in turn led to the development of helpful treatments, but that is not the way it happened at all. Instead, a drug was discovered by accident that proved very helpful in treating schizophrenia. This led to our current understanding of schizophrenia as a chemical disorder in the brain. Here, again, chemical architecture provides the key.

In the late nineteenth century a number of new dyes were synthesized by the German dye industry in the search for better ways to color fabric. One chemical group of these dyes is called the phenothiazines. Earlier, it had been found that some dyes helped in the treatment of malaria. Apparently convinced that these new dyes had to be good for something in medicine as well, physicians tried them on several disorders. Needless to say, this was before any country had developed a Food and Drug Administration. A French surgeon noted in 1949 that the phenothiazine dyes had a marked calming effect on some surgical patients. Soon after, one of these dye compounds, chlorpromazine, was found to have remarkably beneficial effects on schizophrenics. By 1954 it was approved in the United States as a treatment. It is then that the remarkable decline in the number of patients in mental hospitals in the United States began.

Until the introduction of chlorpromazine there were really no helpful treatments for schizophrenia. There were more than two million schizophrenics in the United States, and many of them had to be hospitalized. Chlorpromazine does not cure schizophrenia, but it is effective in treating the more severe symptoms. Patients become calmer and more rational and many are able to live on their own outside the hospital.

The extraordinary success of chlorpromazine led many scientists to think that schizophrenia might be understood at the chemical level by understanding the chemistry of chlorpromazine. However, this hope received a setback when other drugs were discovered that also were very effective in treating the symptoms of schizophrenia. One drug in particular, haloperidol, is even more effective than chlorpromazine, but its chemical structure is quite different. This kind of puzzle is very common in neuroscience. Different drugs having very different chemical structures can have the same effects on the brain and behavior.

When faced with a puzzle like this it is helpful to simplify the situation as much as possible. We cannot, of course, study the chemical activity of the schizophrenic brain directly. But we can look at the effects of drugs on particular neurotransmitter systems in the brain in the laboratory. All mammals' brains have essentially the same chemical neurotransmitter systems. Several of the better-known chemical systems in the brain can be studied in animals; in fact, the appropriate brain tissue can be removed and its chemical reactions studied in a test tube.

It was soon learned that drugs like chlorpromazine and haloperidol both interfered with a chemical transmitter called dopamine in the brain of the rat and other laboratory animals. It is possible to prepare a solution of dopamine or other receptors from the brain of a mammal. We can then measure how well various drugs combine with or bind to these receptors, relative to how well the normal chemical transmitter dopamine binds.

Chlorpromazine, haloperidol, and all other antipsychotic drugs, those that help in the treatment of schizophrenia, bind to dopamine receptors. In fact, haloperidol binds even better than the real transmitter, dopamine. The effectiveness of all the antipsychotic drugs in treating schizophrenia can be predicted quite accurately by measuring how well they can displace haloperidol from the dopamine receptors. The dopamine receptors are treated with radiolabeled haloperidol. Another antipsychotic drug is added and the number of labeled haloperidol molecules it displaces or bumps off the dopamine receptors is measured. This simple measurement predicts almost exactly how effective the given drug will be in treating schizophrenia.

When dopamine is released at synapses onto dopamine-receiving target neurons in the brain, it attaches to the dopamine receptors and activates the target neurons. It causes many changes in the activity and functioning of these neurons, including chemical reactions within the cell and cell membrane, and even on the genetic material, the DNA, in the nucleus of the cell, the second-messenger system described earlier. When the antipsychotic drugs bind to dopamine receptors, they do not activate the dopamine-receiving neurons. The reason they have such powerful effects is because they block dopamine from attaching to the receptors, just as naloxone blocks opiate receptors.

In an old-fashioned door lock with a keyhole on each side, the door can be locked and then a key inserted from the inside. The door cannot now be unlocked with a key from the outside because it is blocked by the key inserted from the inside. So it is with the antipsychotic drugs. They attach to dopamine receptors and literally block dopamine from attaching to the receptors.

Chlorpromazine and haloperidol have different chemical structures. When chemists speak of the structure of a molecule, they refer to the particular chemical features—for example, how many carbon atoms, how many nitrogen atoms, presence of a benzene ring or side chains, and so on. At this level chlorpromazine and haloperidol are indeed quite different. However, the *shape* of a part of the chlorpromazine molecule is similar to the shape of a part of the haloperidol molecule, a shape very like that of the dopamine molecule.

This is the ultimate architecture of the brain. The receptor molecule is the lock that can be opened only by the transmitter chemical that has the right shape, the key molecule. However, many other chemicals can also have the same key shape on one part of the molecule. So it is with drugs that attach to chemical receptors in the brain. The complete molecule of the drug may be very different chemically from the normal transmitter molecule, but one part of it will have a shape that is similar to the shape of the transmitter molecule. It is this architectural structure of the molecule that determines its action on chemical receptors in the brain. Many drugs that act on the brain are extracted from plants and have chemical structures quite different from normal brain chemicals. But, by chance, a part of the molecule of the plant drug has an architecture similar enough to the brain chemical to fool the chemical receptor in the brain.

LSD (lysergic acid diethylamide), the drug that causes profound and bizarre psychoticlike effects, is extremely potent—a few millionths of a gram is an effective dose. It is so powerful because it happens to have the right shape to act on a particular chemical receptor in the brain. You might think that because LSD produces psychoticlike symptoms—delusions and hallucinations—it would act on the dopamine receptor system, but such is not the case. It apparently acts on a different chemical transmitter system in the brain, the serotonin system. Little is known about the functions of the serotonin system at present beyond the fact that it seems to play an important role in the regulation of body temperature and sleep and waking.

The extraordinary fact that all drugs effective in treating schizophrenia block the dopamine receptor, and are effective in proportion to how much they block it, would seem to imply that schizophrenia is caused by too much dopamine. This was the initial dopamine theory of schizophrenia.

There have been several studies in which the brain levels of dopamine have been measured in schizophrenic patients who died. The results are negative—the brain level appears to be normal. There is some very recent work indicating that such patients do have a significant increase in the number of dopamine receptors in the brain. If this holds up, it would fit beautifully with the chemical studies. The schizophrenic brain would then be much more sensitive to dopamine than the normal brain. So it would not be that there is too much dopamine but rather that the normal amount of dopamine has too powerful an action. By blocking dopamine receptors with antipsychotic drugs, the system can be brought back toward a normal level of sensitivity and function.

The purely accidental discovery that a certain type of dye molecule can help treat the symptoms of schizophrenia has led to real advances in our understanding of schizophrenia and its treatment. The cure may be a long time in coming, but improved treatments are at hand.

A very different kind of disorder, Parkinson's disease, also involves the dopamine chemical messenger in the brain. Let's look at a typical (fictional) case history:

James was a very successful engineer. In his mid-fifties he noticed that his hands began to tremble a little, even though he was known for his steady hand. He also realized that it was more difficult to begin a movement. When the doorbell would ring, it seemed to require a real effort to get up from his chair. At first he thought he might have injured some muscles, but his condition gradually became worse. He walked more slowly, with a shuffling gait, and the tremor movement of his hands increased. However, his intellectual abilities and thought processes remained normal. James had developed Parkinson's disease.

Parkinson's disease is a progressive motor disorder. Its incidence is about one in one thousand in the total population, but it is much more frequent than that in older people. Most readers have seen elderly people with the disease. They have a slow, shuffling walk, are stooped over, and may show repetitive movements such as "pill rolling" with the fingers. The major symptom is difficulty in starting and sustaining voluntary movements. Many people with the disease are able to compensate remarkably for the problems it creates. Modern treatments have provided much relief for the symptoms. Indeed, these treatments are one of the success stories of basic research in neuroscience.

It has been known for a long time that Parkinson's disease involves abnormalities in a particular part of the brain. There is a structure in the lower part of the brain called the substantia nigra ("dark substance"), so called because the nerve cells contain a dark-colored pigment. Many nerve cells in the substantia nigra send their axons to a higher brain structure called the caudate nucleus, which lies in the forebrain buried beneath the cerebral cortex. The caudate nucleus has been known for some time to be involved in the regulation of bodily movements.

In 1955 a German anatomist found that there were significantly fewer neurons in the substantia nigra in the brains of deceased persons who had suffered from Parkinson's disease. The next breakthrough came with the development in Sweden of an extraordinary technique to visualize certain neurons in the brain. Neurons that contain the right transmitter chemicals will shine or fluoresce if they are treated with a chemical called formaldehyde (the ingredient in embalming fluid). The neurochemical dopamine fluoresces with a particular color. Using this method, it was found that the nerve cells in the substantia nigra contain a great deal of dopamine, the same transmitter chemical impli-

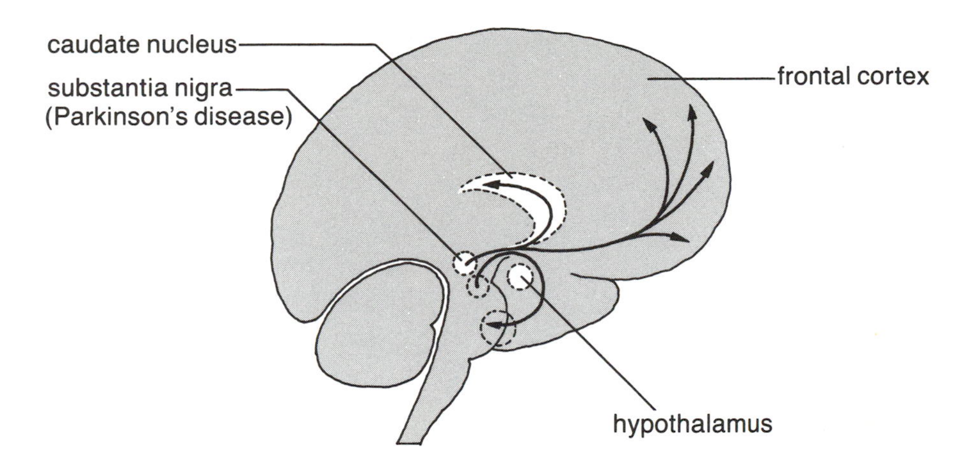

Dopamine Pathways from the Midbrain

cated in schizophrenia. In deceased persons who suffered from Parkinson's disease the dopamine neurons in the substantia nigra are very much fewer in number and lower in dopamine content. It seemed clear that Parkinson's disease was due to a loss of these dopamine-containing neurons in the substantia nigra.

The treatment involves a drug called L-dopa that these neurons can make into dopamine. The few remaining dopamine neurons in the substantia nigra can make more dopamine and hence yield more normal function. Administration of L-dopa produced dramatic and rapid improvement in many Parkinson's patients:

> The effect of a single intravenous administration of L-dopa was, in short, a complete abolition or substantial reduction of akinesia [inability to initiate movements]. Bedridden patients who were unable to sit up, patients who could not stand up from a sitting position, and patients who when standing could not start walking performed all these activities with ease after L-dopa. They walked around with normal associated movements and they could even run and jump. The voiceless, aphonic speech, blurred by unclear articulation, became forceful and clear again as in a normal person. For short periods of time the patients were able to perform motor activities which could not be prompted by any other known drug to any comparable degree. This L-dopa effect reached its peak within 2 to 3 hours and lasted, in diminishing intensity, for 24 hours. [From the first clinical report on

L-dopa treatment: O. Hornykiewicz. The mechanisms of action of L-dopa in Parkinson's disease. *Life Sciences* 15 (1974), 1249–1259.]

It may have occurred to you that some of the signs of old age in normal people resemble a very mild form of Parkinson's disease. Very elderly people tend to be slow moving and slower to initiate movements. They may even have very minimal tremors. There is no evidence as yet of any significant cell loss in the substantia nigra in normal old people. However, we might speculate that there is a decrease in dopamine in the substantia nigra–caudate system as a normal consequence of old age.

Old rats are very much like old people. They are slow moving and do not initiate movements as well as young rats. Indeed, the aged rat has become widely used as a general model for old age in mammals, including humans.

A rather striking test of the ability to initiate and sustain movement in the rat is the swimming challenge. A rat is placed in a glass bell jar partly filled with water so that the animal cannot climb out. A normal young rat will swim vigorously for some time until rescued by the experimenter. An aged rat will just lie in the water, sink, and drown if not rescued immediately. Administration of L-dopa to aged rats results in vigorous swimming behavior—they act like young rats. We hasten to add that L-dopa has not yet been administered to normal old people. However, these experiments on old rats do raise the hope that it may someday be possible to reverse some of the symptoms of old age in humans.

We saw that all the drugs effective in treating the symptoms of schizophrenia block dopamine receptors in the brain. This suggested that schizophrenia might be due to too much dopamine or perhaps too many dopamine receptors active in the brain—in other words, just the opposite to the cause of Parkinson's disease, too little dopamine.

Both of these ideas seem clear enough, but they don't seem to fit together at all. Why should the simple motor pathway involved in Parkinson's disease have anything to do with the complex mental symptoms of schizophrenia? The answer is it probably doesn't. There are two major dopamine circuits in the brain. One is the substantia nigra–caudate pathway involved in Parkinson's disease. The other is a much more widespread and diffuse system with dopamine-containing cell bodies in the brainstem that send their axons up to the highest regions of the brain—the cerebral cortex and the limbic system—the brain systems concerned with thought and consciousness, the complex higher mental functions. This dopamine system is also a one-neuron pathway, but the dopamine cell bodies are located in clusters in the brainstem rather than in the substantia nigra and send their axons in a very extensive array to neurons in the higher brain regions.

These two dopamine systems do not seem to be related to each other in terms of functions. They make use of the same chemical transmitter—dopamine—but seem to be involved in very different matters, one concerning movement control and the other concerning intellectual functions. This illustrates a general point about chemical transmitters in the brain. The chemical transmitter molecules are indeed the messengers, but they are not the message itself. The movement-control pathway from the substantia nigra to the caudate nucleus makes use of dopamine as its messenger to convey information at synapses in the caudate nucleus. Dopamine simply transmits this information. The reason this is information about movements is that the caudate nucleus is connected with those parts of the brain concerned with movement control. Dopamine itself carries no message other than to act as a synaptic transmitter. The other dopamine system that projects to the higher brain regions is involved in intellectual functions because these brain regions are. The dopamine carries no such information itself.

What about the effects on movements of the drugs like chlorpromazine and haloperidol that are used to treat schizophrenia? They cause a decrease in dopamine actions in the brain. By the same argument they ought to cause movement disorders like Parkinson's disease. This did not appear to be the case until recently. A new disease has appeared among long-term patients in mental hospitals called tardive dyskinesia (TD). It begins with small, involuntary movements of the face and can progress to the point where the patient makes uncontrollable grimacing movements. Even in the early stages it can pose special problems for a schizophrenic patient. Schizophrenia is hard enough to deal with; adding uncontrolled and weird movements of the face makes life even more difficult. Estimates vary, but a significant number of patients who have been on antipsychotic drugs for years, perhaps 25 to 40 percent, develop TD.

In the early stages of TD, the symptoms can be reversed by taking the patient off antipsychotic drugs. Of course, this means that the patient's schizophrenic symptoms may get worse—a double bind for the patient and for the psychiatrist.

The reason TD was not seen earlier is that it typically takes years of treatment with an antipsychotic drug before it develops. It is a striking and most unfortunate example of what is called an iatrogenic disease—a disease caused by a treatment. The movement disorder is a slowly developing side effect owing presumably to a long-term and permanent alteration in the dopamine system concerned with movement control.

We began our grand tour of the brain by strolling through the ramshackle house that nature built into the most remarkable and complex structure we know of. We saw how the major parts of the brain are put together in what seemed a rather senseless way until we looked at how the brain developed and evolved over the eons of life on earth.

We then made use of the "incredible shrinking man" trick and entered one of the major rooms in the house, the visual region of the cerebral cortex. It turned out to be not just one room but a whole series of rooms, each seemingly specialized for one or another aspect of visual experience. All of the rooms were composed of the six layers of neurons that make up the cerebral cortex, but there was more: each room was filled with columns of nerve cells sensitive to the finer-grained features of seeing. The column of nerve cells in the cortex seems to be the ultimate unit of experience in the visual cortex and also for the auditory and somatic (body) senses. The column may even be the ultimate experiential unit in the still-mysterious association areas of the cerebral cortex.

Shrinking further, we came upon the basic building blocks of the brain, the neurons. The nerve cell makes special use of the little ion channels through its membrane to send a message, the nerve impulse, out its axon to other neurons. When the nerve impulse reaches the synapse, the functional connection of the axon with another neuron, a very different kind of messenger takes over. Shrinking still further, we saw that this chemical messenger consists of molecules that are released by the terminal at the synapse and attach to receptor molecules on the target neuron to excite or inhibit it.

The chemical transmitter molecules and the receptor molecules on nerve cells that they attach to seem to hold some of the ultimate secrets of how the brain works, as in how we experience pain and pleasure, and how the brain might become disordered in mental illness.

The chemical molecules are indeed the messengers, but they are not the messages. The immensely complex and highly structured circuitry of the brain is the ultimate repository of the mind. In the next chapters we look at some of the more significant and often astonishing things the brain can do.

A Young Man Recognizes His Mother

or

A simplified tour of the visual system for the busy reader

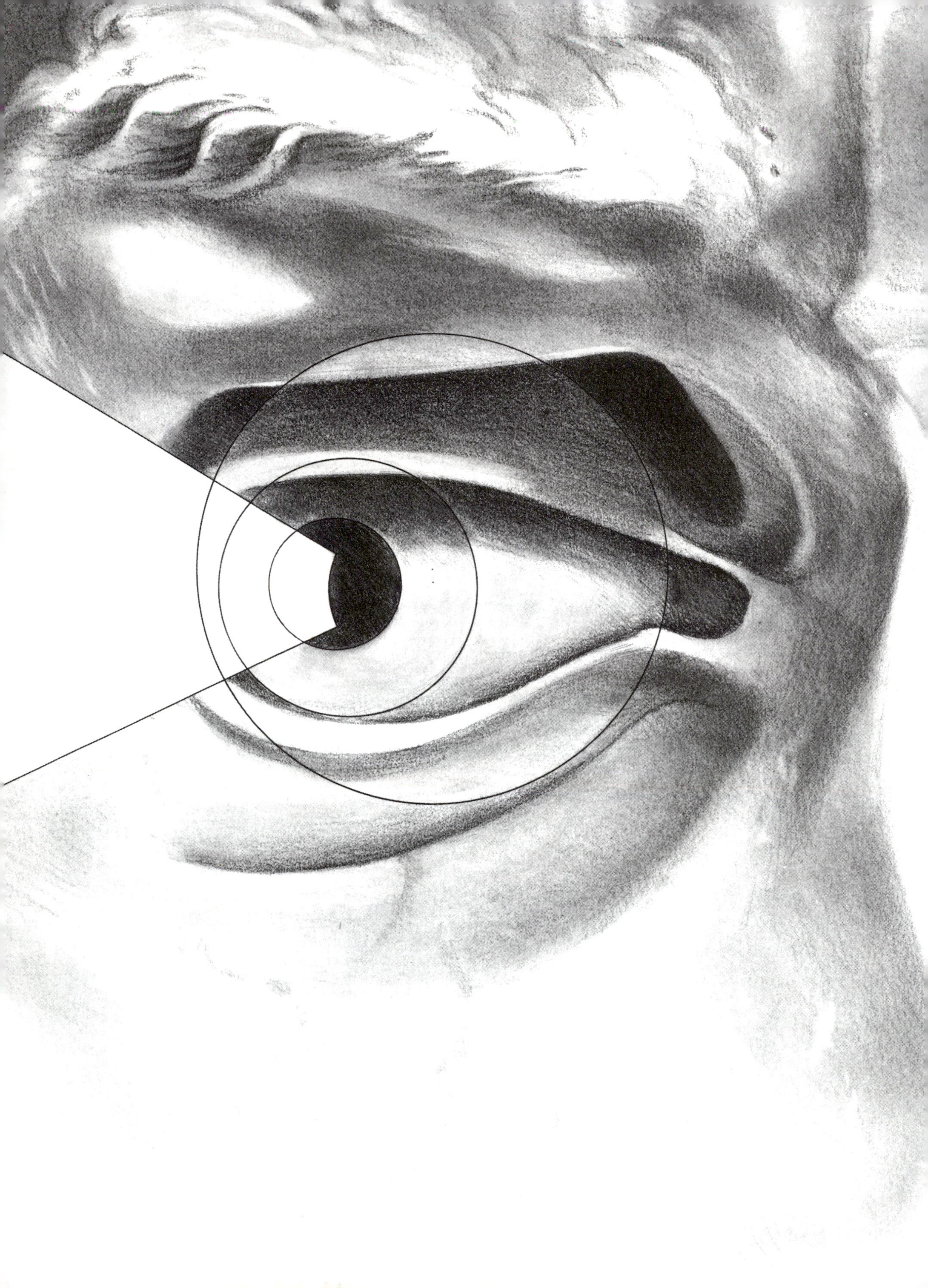

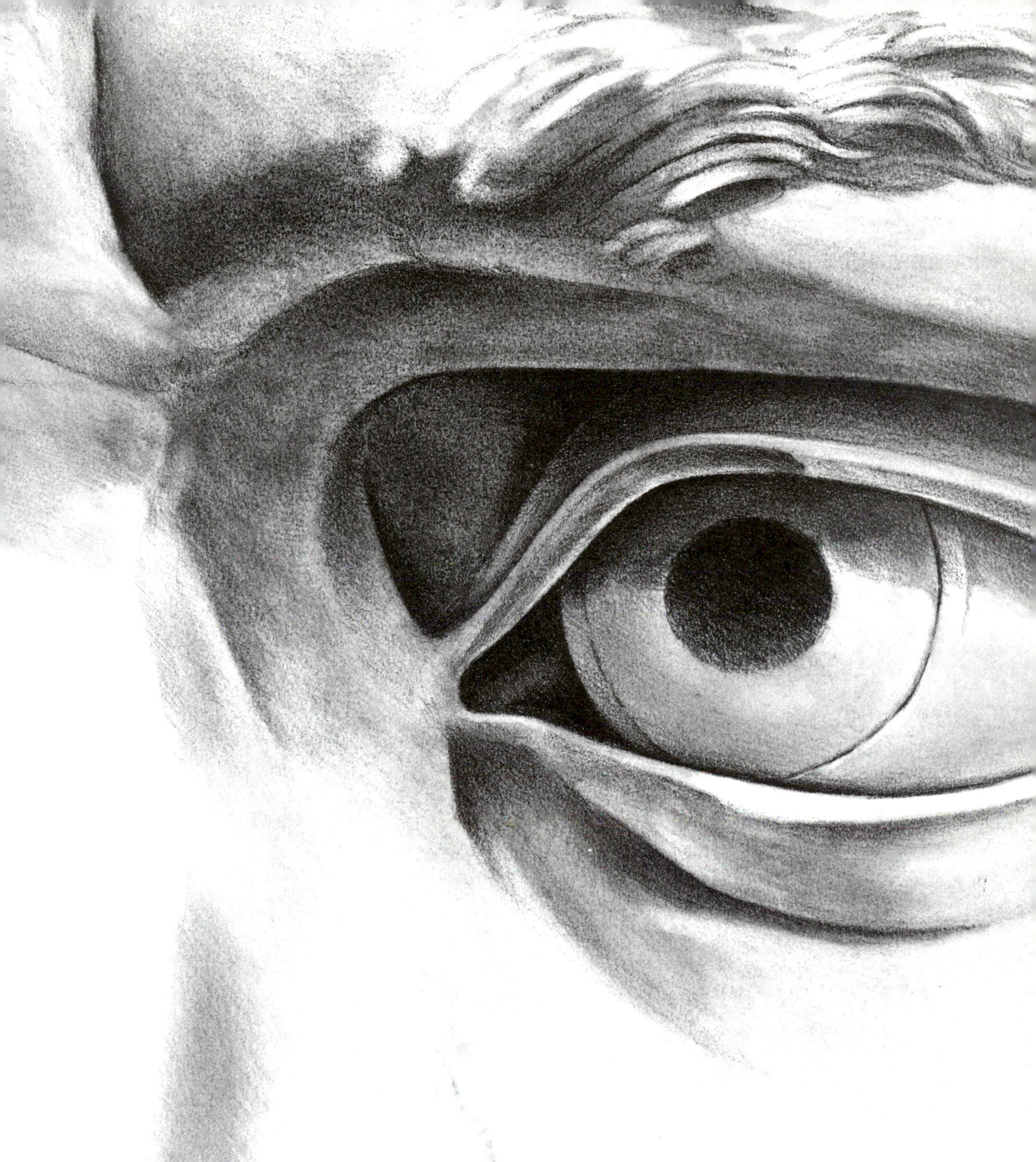

David stands on a high marble plinth. Every day between 9 and 2 (9 and 1 on Sundays and festival days) many people come to see him. Sometimes his mother comes to see him at work. When she enters his field of vision, wonderful things happen inside his famous head. Her image penetrates David's eyes through an opening in the front of each one, called the pupil.

Immediately behind each pupil is a lens that focuses and projects the image on a light-sensitive layer of nerve cells called the retina, which lines the inside of the back of the eyeball. Since the image is inverted as it passes through the lens, you will have to turn the book upside-down to continue reading.

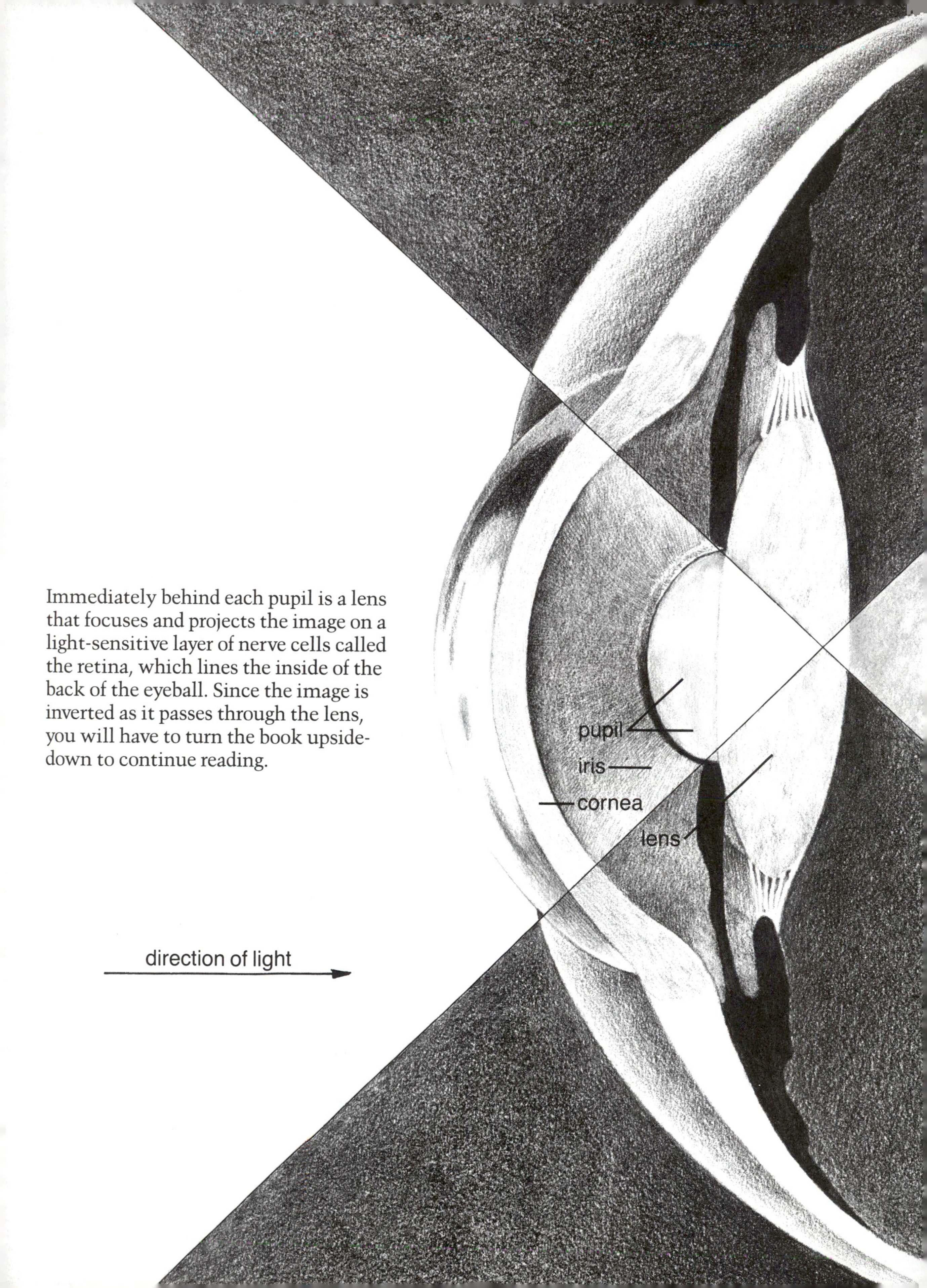

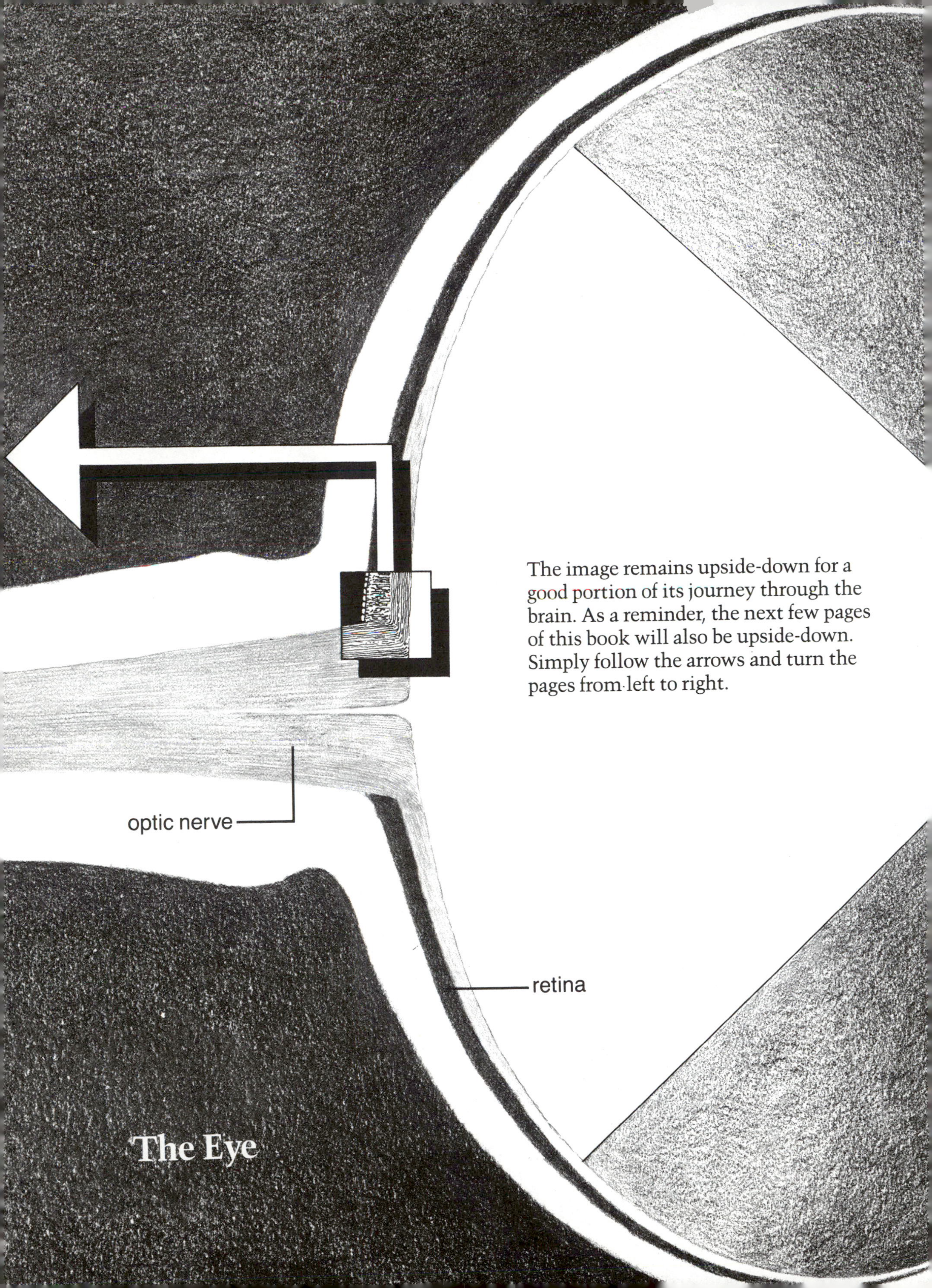
The image remains upside-down for a good portion of its journey through the brain. As a reminder, the next few pages of this book will also be upside-down. Simply follow the arrows and turn the pages from left to right.
optic nerve
retina
The Eye

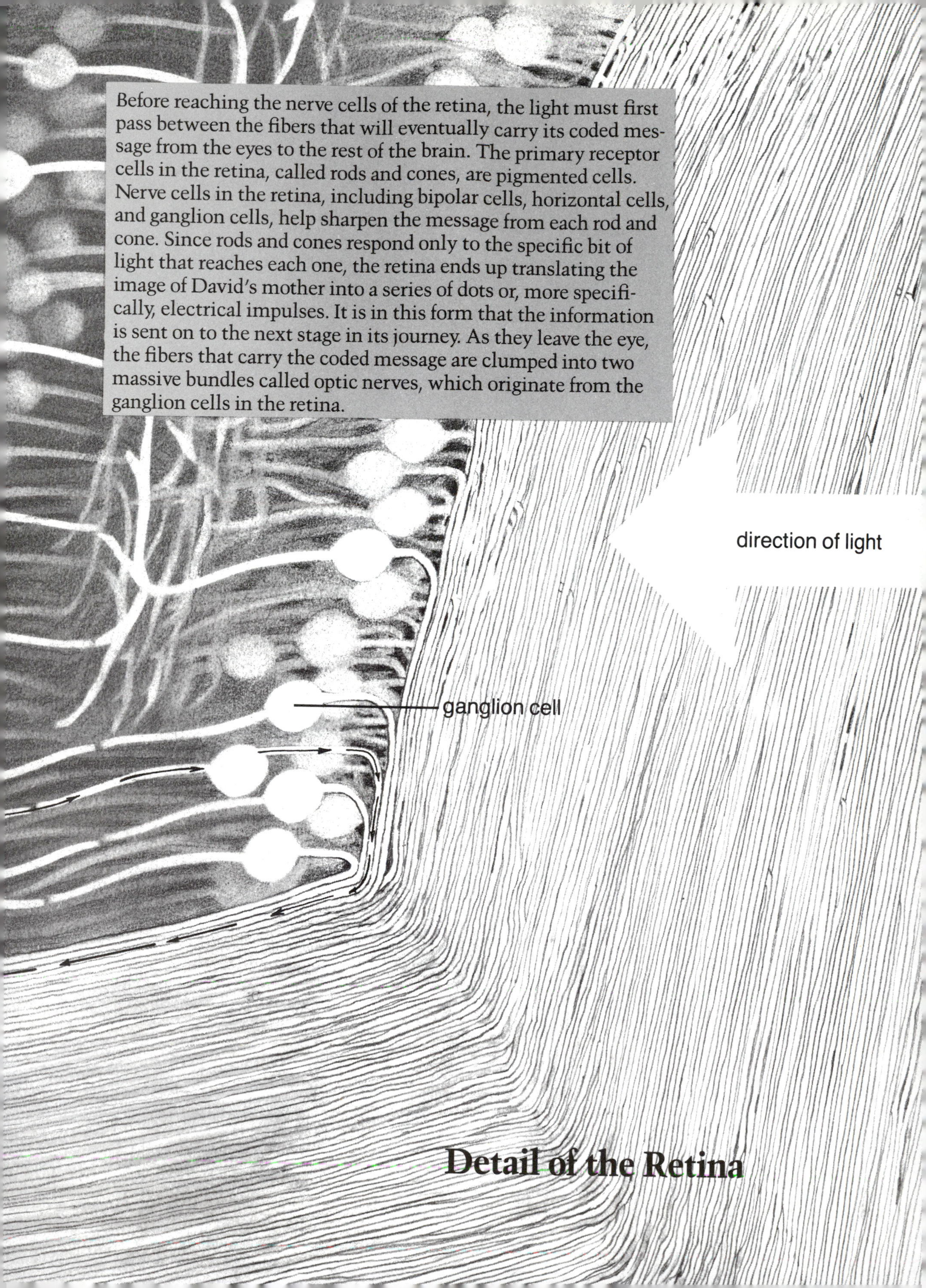

Before reaching the nerve cells of the retina, the light must first pass between the fibers that will eventually carry its coded message from the eyes to the rest of the brain. The primary receptor cells in the retina, called rods and cones, are pigmented cells. Nerve cells in the retina, including bipolar cells, horizontal cells, and ganglion cells, help sharpen the message from each rod and cone. Since rods and cones respond only to the specific bit of light that reaches each one, the retina ends up translating the image of David's mother into a series of dots or, more specifically, electrical impulses. It is in this form that the information is sent on to the next stage in its journey. As they leave the eye, the fibers that carry the coded message are clumped into two massive bundles called optic nerves, which originate from the ganglion cells in the retina.

Detail of the Retina

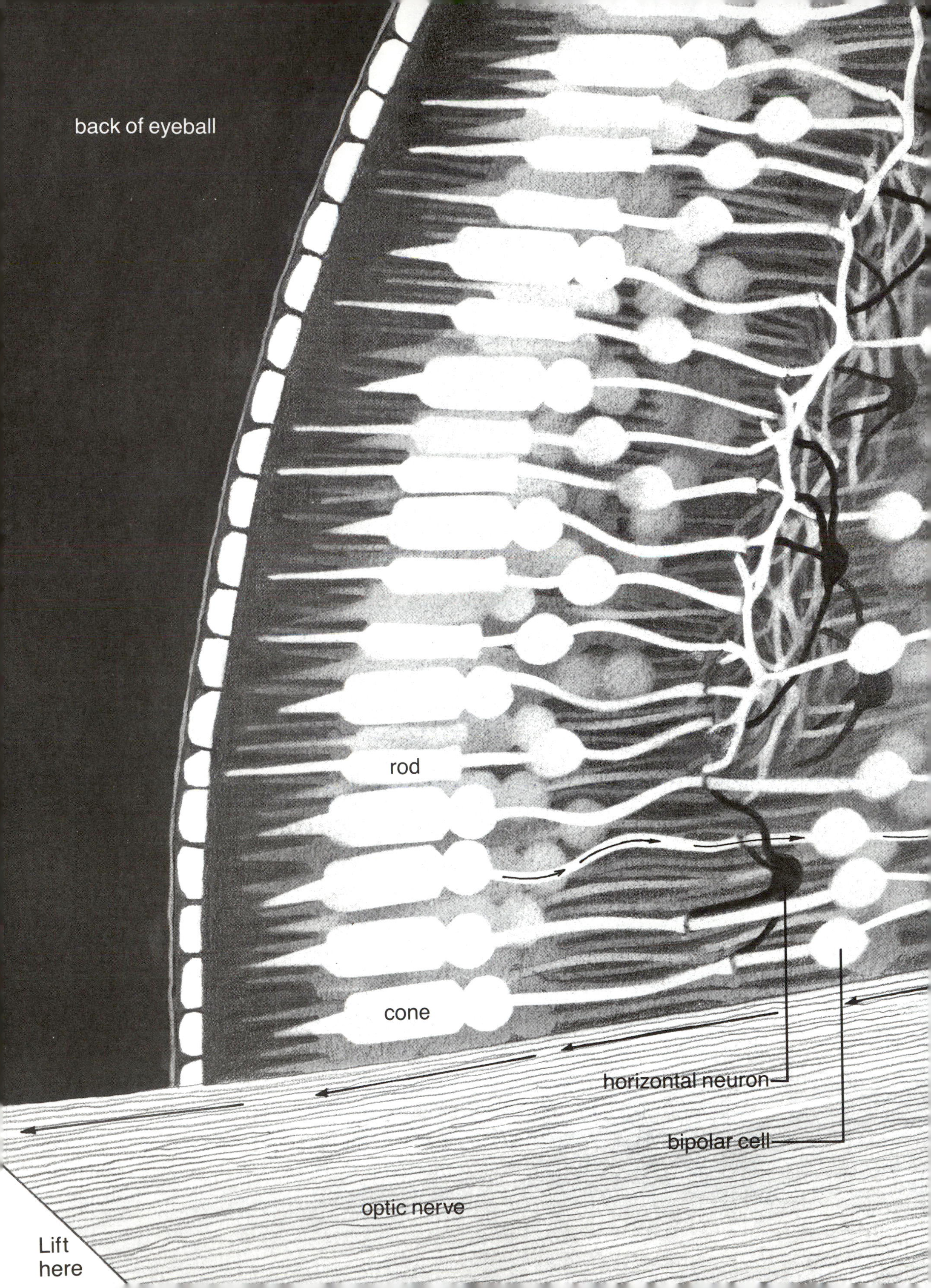
back of eyeball
rod
cone
horizontal neuron
bipolar cell
optic nerve
Lift here

The Primary Visual System

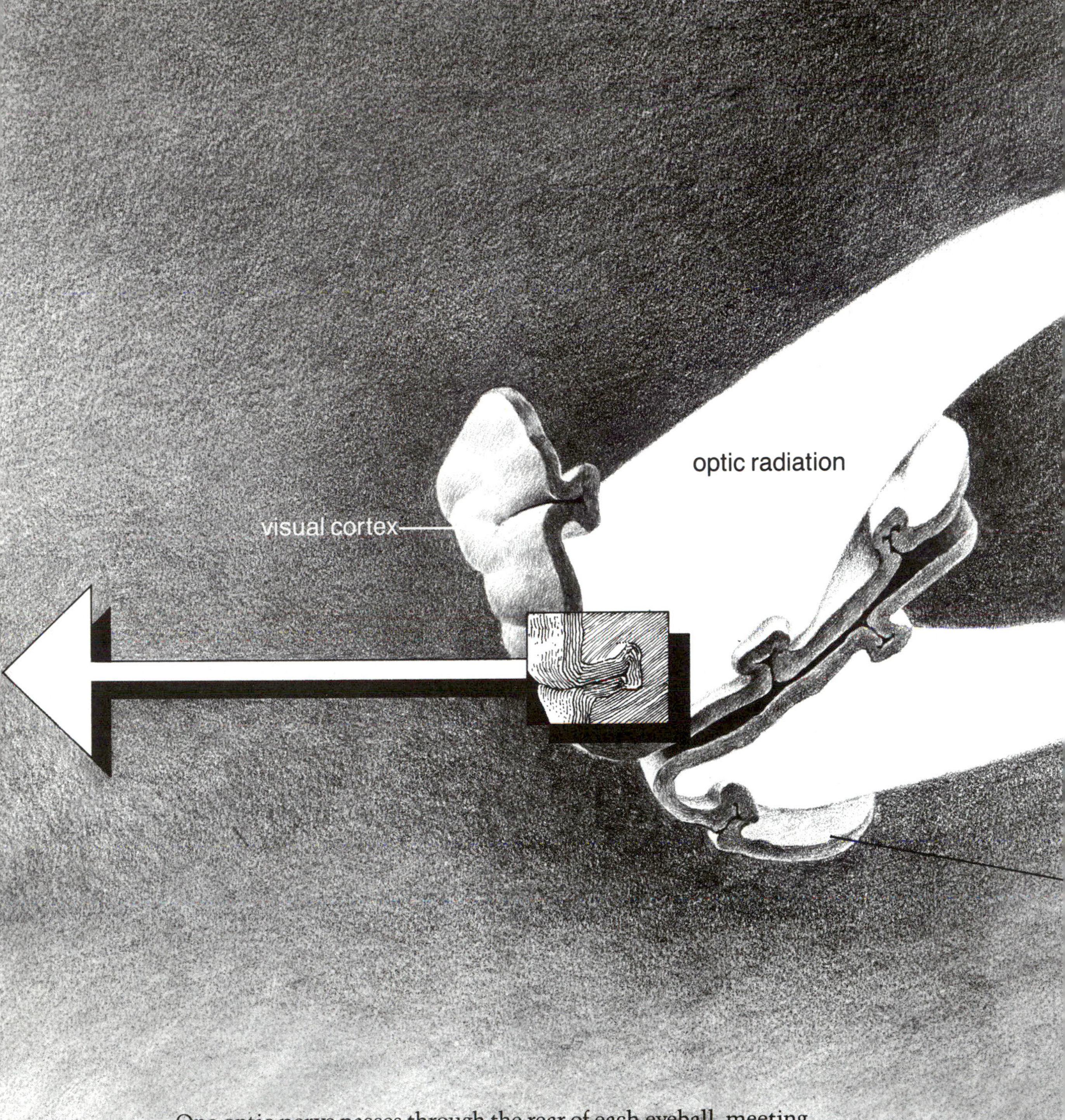

One optic nerve passes through the rear of each eyeball, meeting shortly thereafter at the optic chiasm. Here, about half the fibers from each eye cross over to the opposite side of the brain while the other half continue on the same side. The two rearranged bundles of fibers leaving the optic chiasm are called optic tracts, and they deliver the message to an area of each thalamus called the lateral geniculate body. Although each electrical impulse registers in a particular layer of the lateral geniculate body, the complete message remains unchanged and continues on through two large sets of fibers, called the optic radiation, to the occipital lobe, or visual cortex, of each hemisphere for analysis.

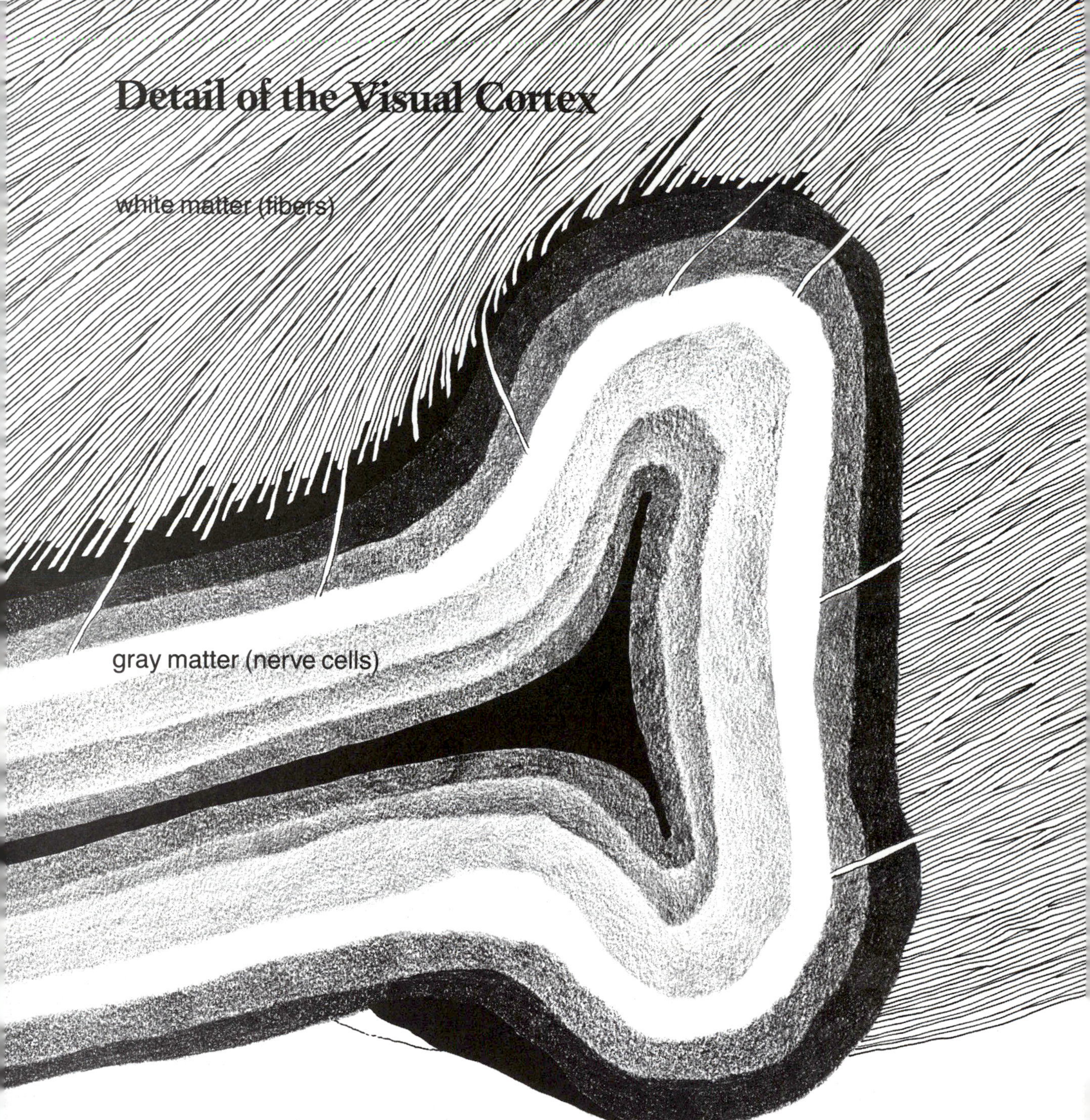

A cross section of the visual cortex shows its six-layered construction. Each layer contains cells of certain shapes and complexity that specialize in responding to different kinds of information. They interconnect within each layer as well as being organized in tiny columns, which extend from layer I through to layer VI. All information from the eyes first enters layer IV and from there passes to the other layers of the visual cortex.

The black and white stripes on the surface of the cortex represent the tendency of each eye to dominate alternating areas of cells. This is called ocular dominance. Cells dominated by the right eye are represented by the white stripes, and those dominated by the left eye fall in the black stripes. This arrangement is the result of the original competition for cells of the visual cortex that went on between David's eyes while he was just a little chip off the old block.

I
II
III
IV
V
VI
cells dominated
by left eye
cells dominated
by right eye

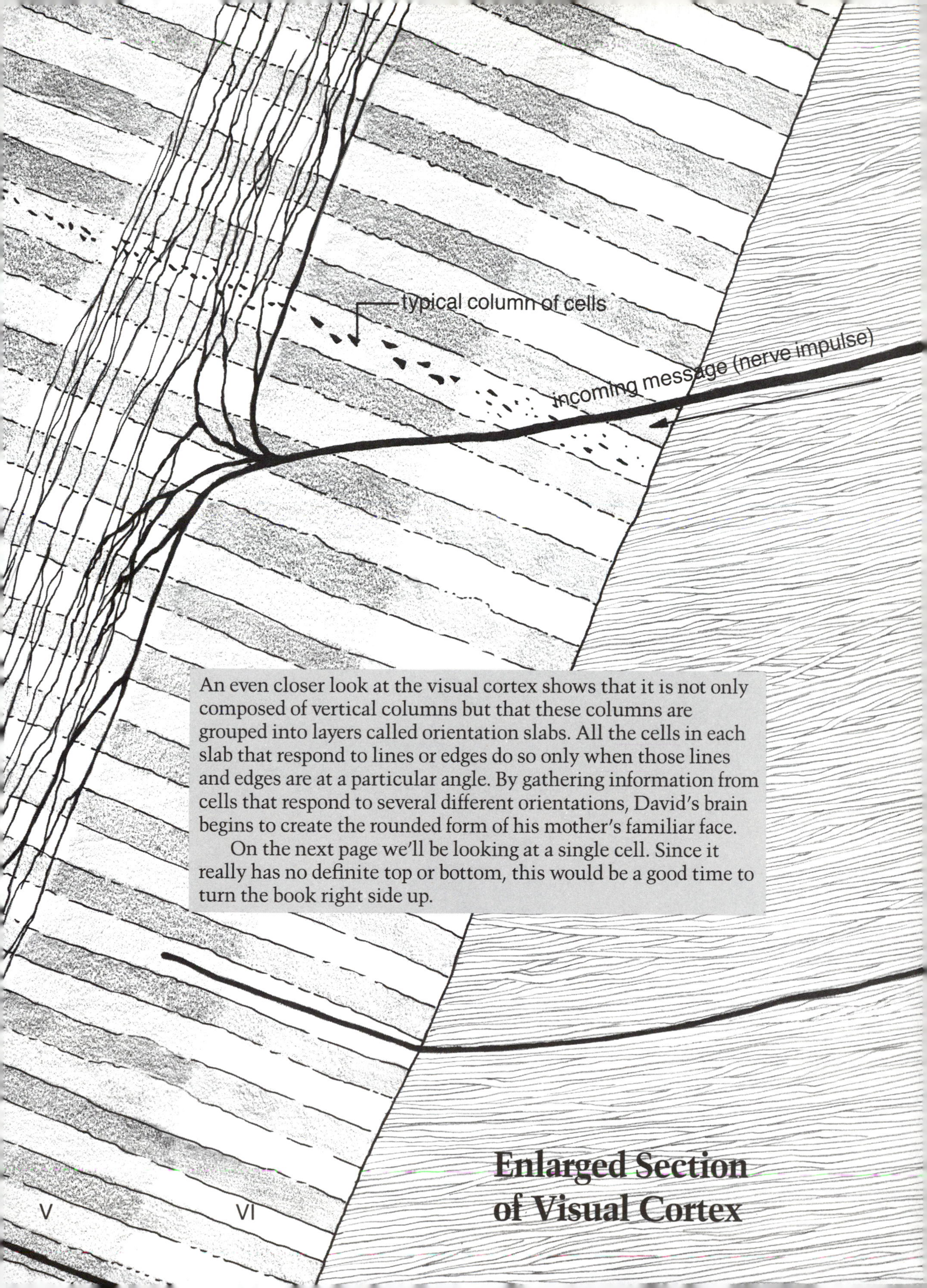

An even closer look at the visual cortex shows that it is not only composed of vertical columns but that these columns are grouped into layers called orientation slabs. All the cells in each slab that respond to lines or edges do so only when those lines and edges are at a particular angle. By gathering information from cells that respond to several different orientations, David's brain begins to create the rounded form of his mother's familiar face.

On the next page we'll be looking at a single cell. Since it really has no definite top or bottom, this would be a good time to turn the book right side up.

Enlarged Section of Visual Cortex

orientation slabs
outgoing message
I
II
III
IV

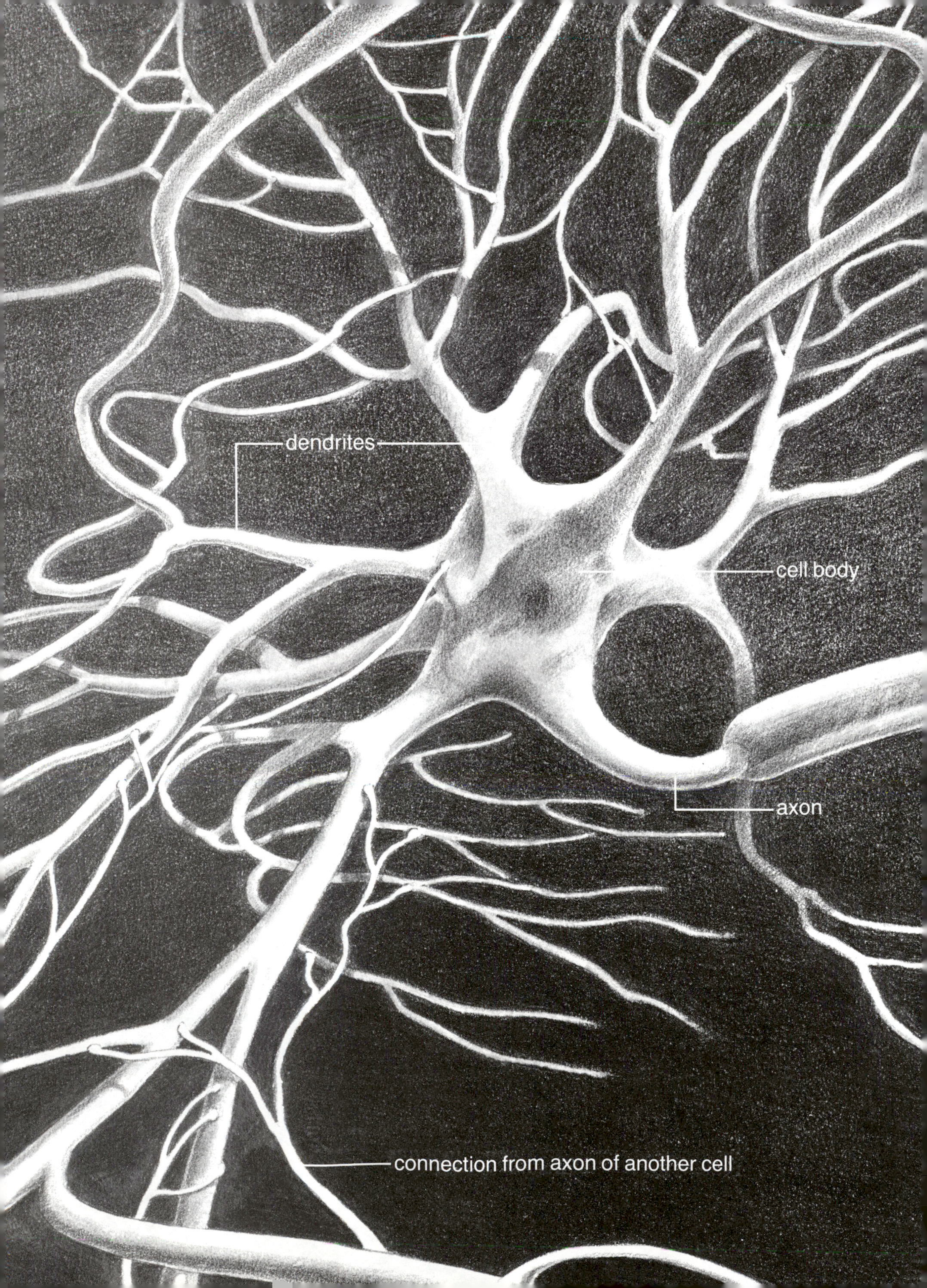
dendrites
cell body
axon
connection from axon of another cell

A Typical Cell

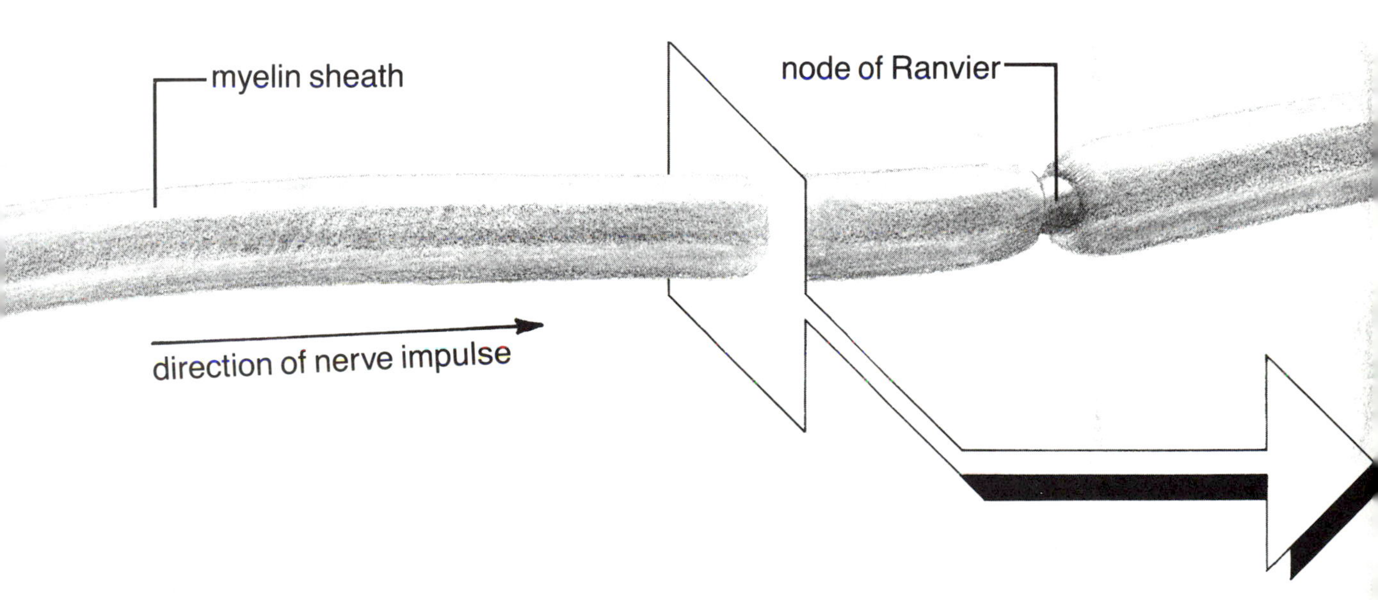

The cortex is made up of billions of cells with billions of interconnections. Each cell body sends out many branches. The largest branch or fiber is called the axon, and it carries the message from the cell body. All the other branches are called dendrites, and they receive messages from the axons of other cells.

The skin, or membrane, that encloses each nerve cell contains tiny holes called channels, which allow certain molecules to pass through. The message or nerve impulse is carried along the axon by the sequential movement of electrically charged particles (called ions) through the channels.

myelin sheath
axon membrane

When those channels nearest the cell body open, allowing ions to enter the axon, the nerve impulse has begun. The influx of ions reverses for a very short time the electrical balance inside and outside the axon at that particular location. By the time the original electrical balance is restored, the channels immediately adjacent have opened and the nerve impulse has moved to that new location. This chain reaction continues all the way to the end of the axon, taking the message with it.

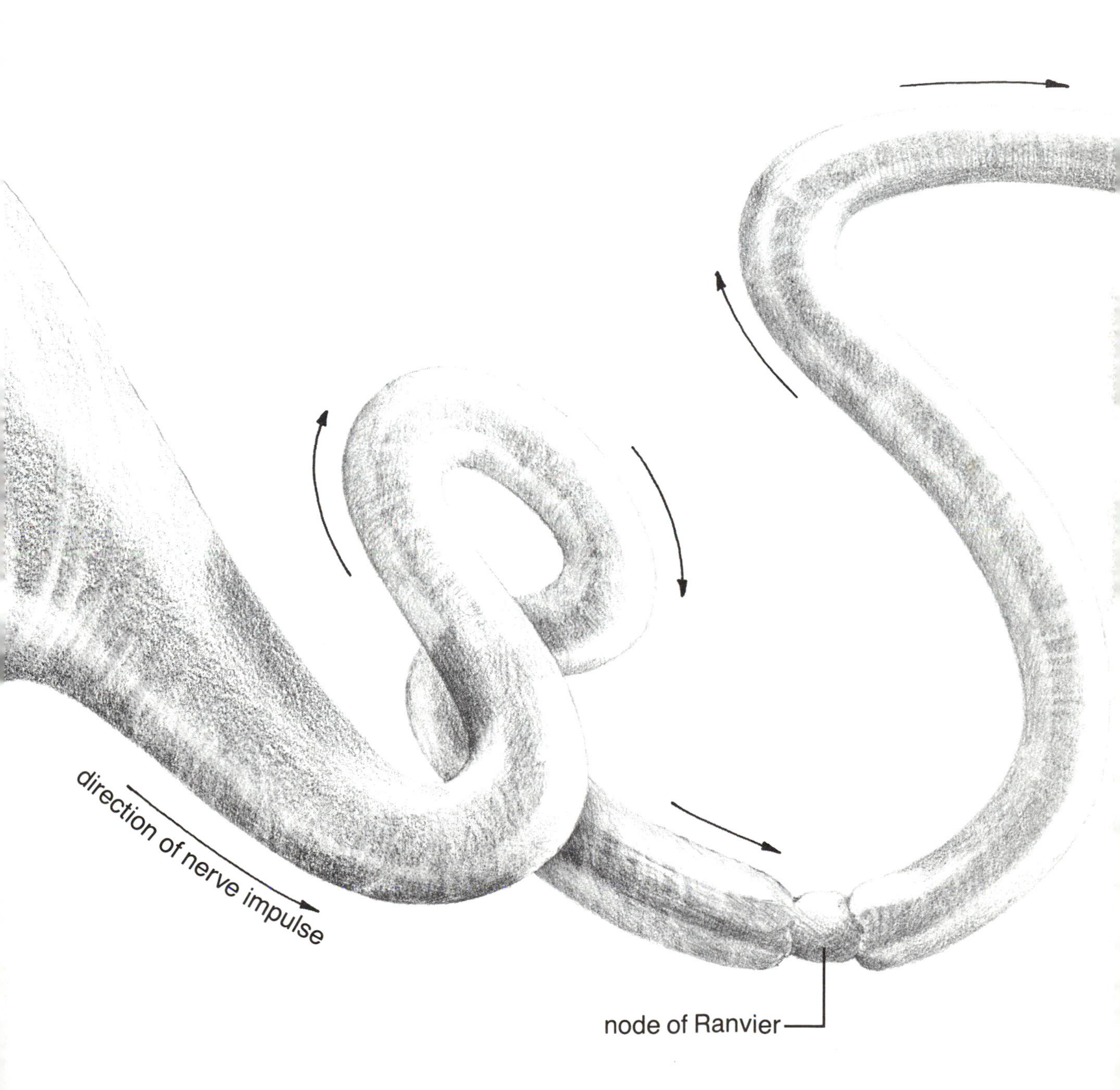

When an axon is covered with a myelin sheath, which acts as a kind of insulation, the influx of ions is limited to those areas between each segment of myelin called the nodes of Ranvier. Because of this, the nerve impulse travels much more quickly down a myelinated axon. This is a good thing when we realize that some axons in the human body are over three feet long. When the nerve impulse reaches the end of the axon, it sets off a reaction in the many synaptic buttons that connect to the dendrites of the receptor cells.

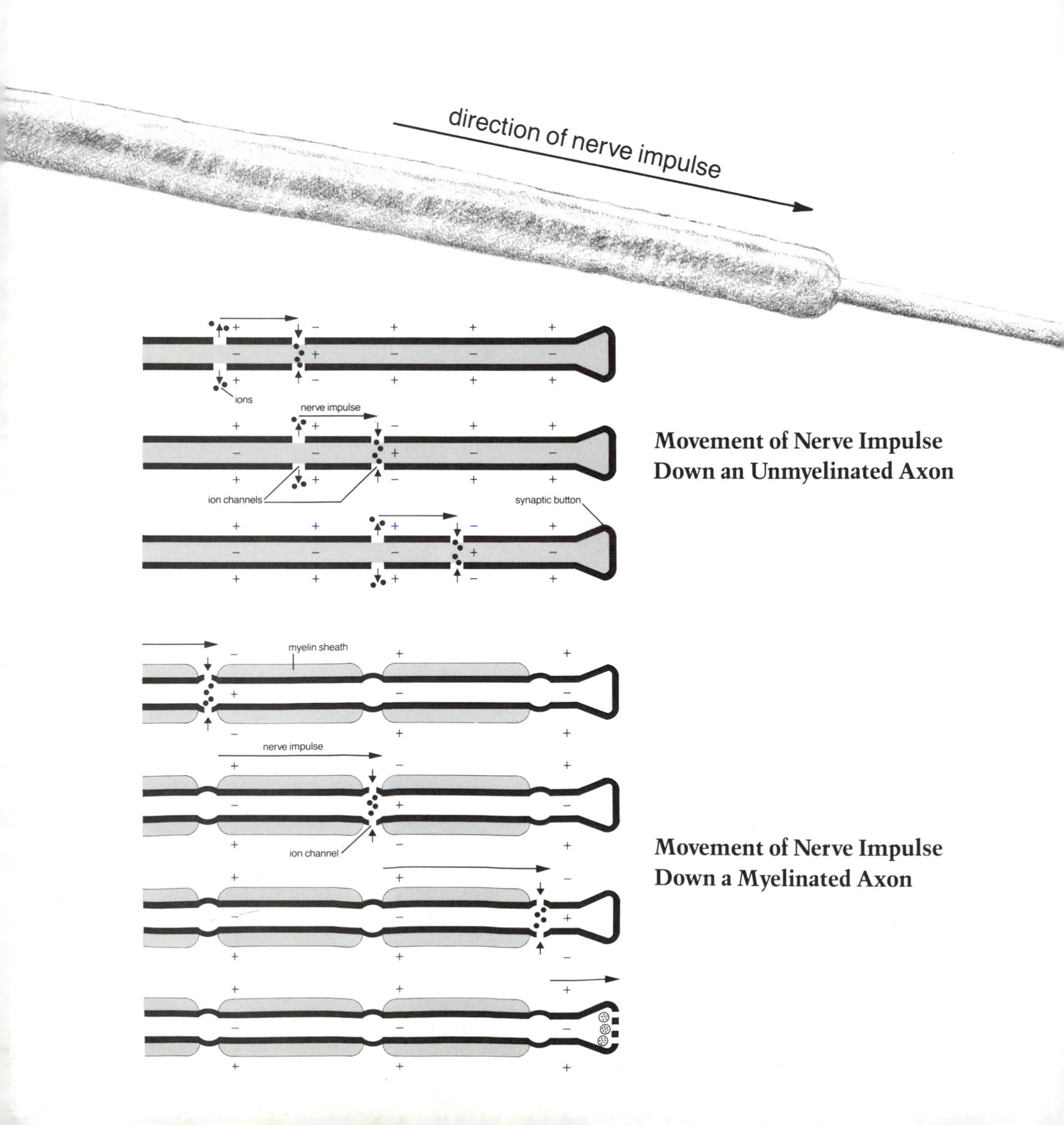

Movement of Nerve Impulse Down an Unmyelinated Axon

Movement of Nerve Impulse Down a Myelinated Axon

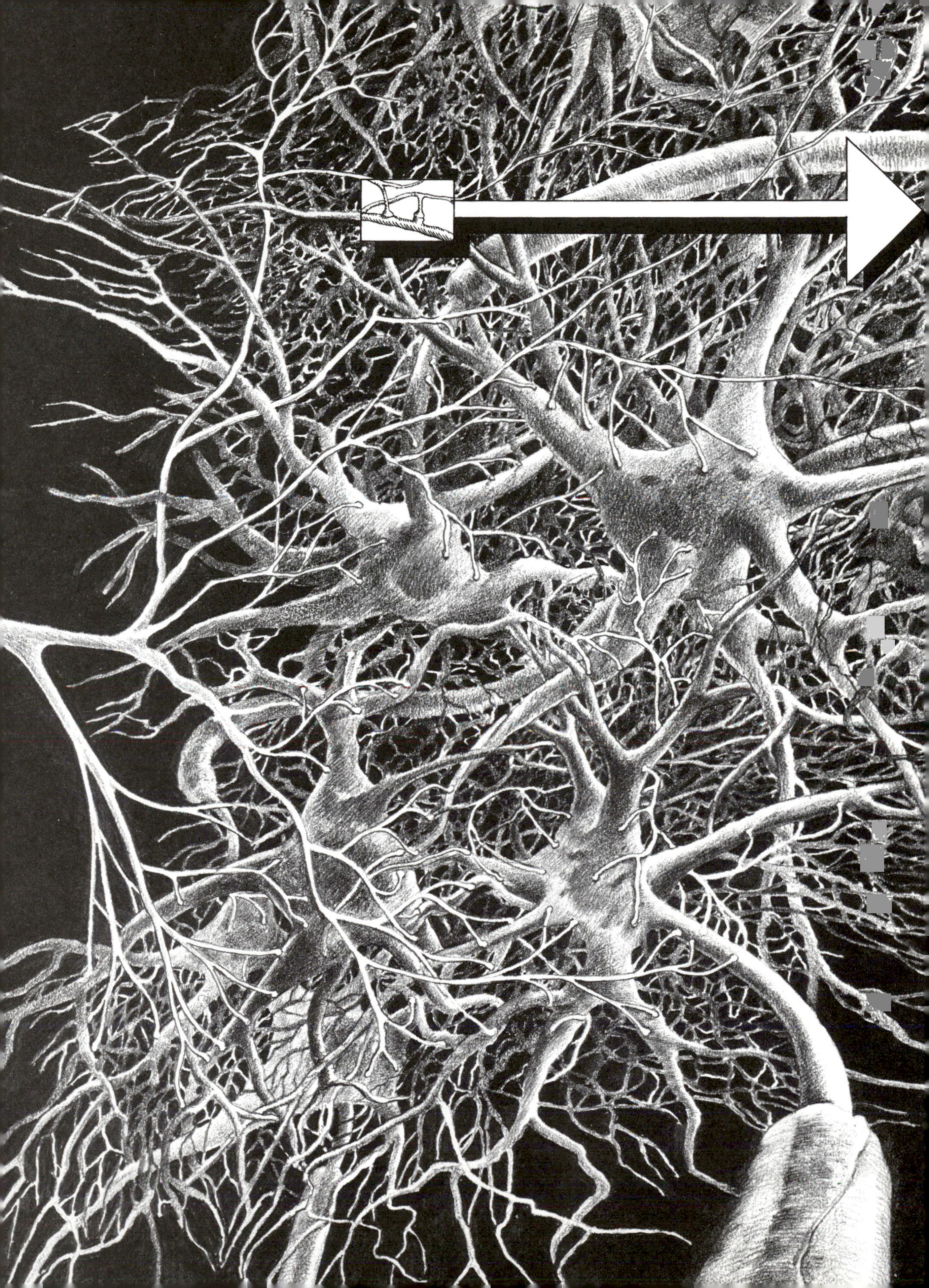

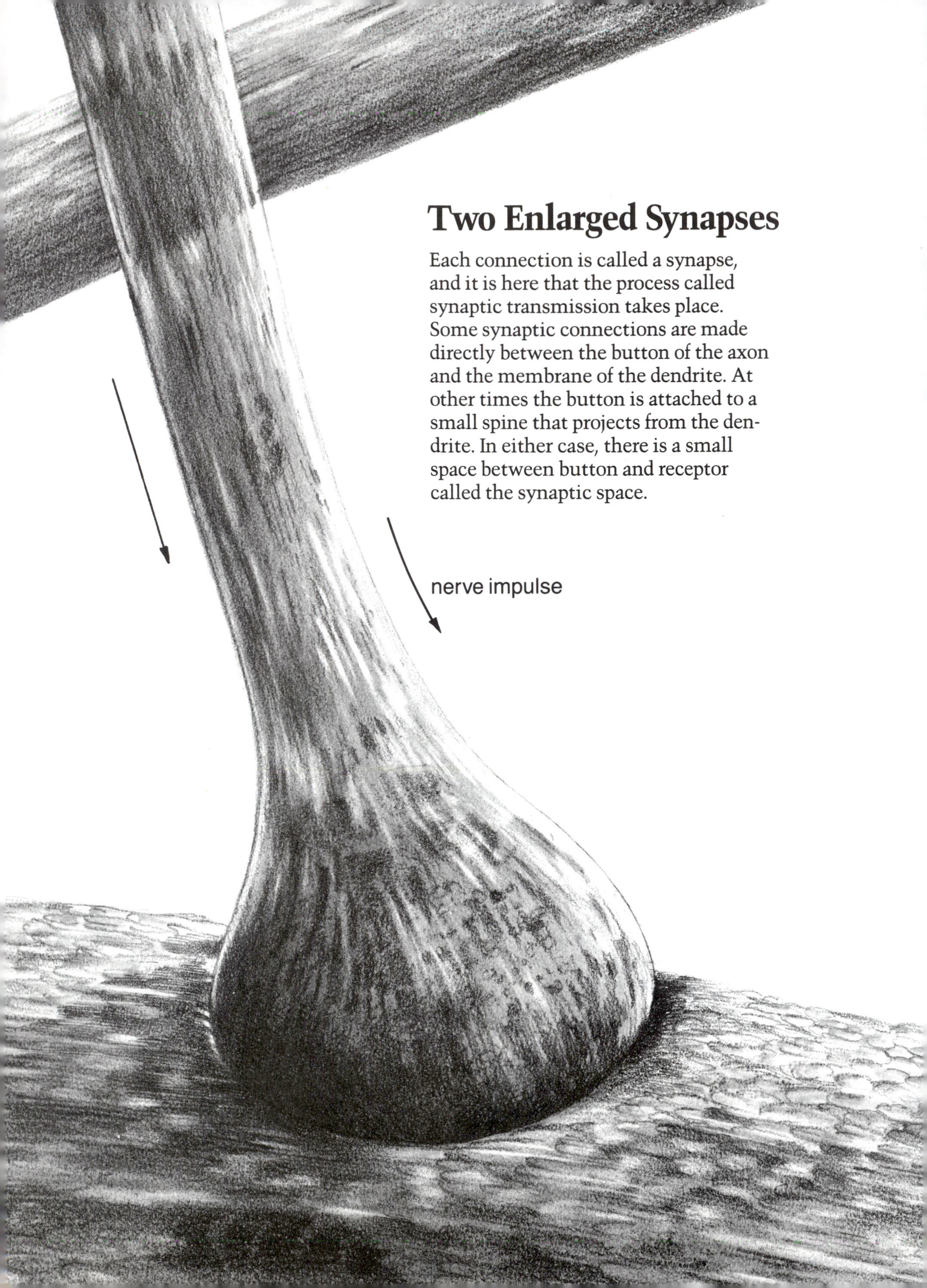

Two Enlarged Synapses

Each connection is called a synapse, and it is here that the process called synaptic transmission takes place. Some synaptic connections are made directly between the button of the axon and the membrane of the dendrite. At other times the button is attached to a small spine that projects from the dendrite. In either case, there is a small space between button and receptor called the synaptic space.

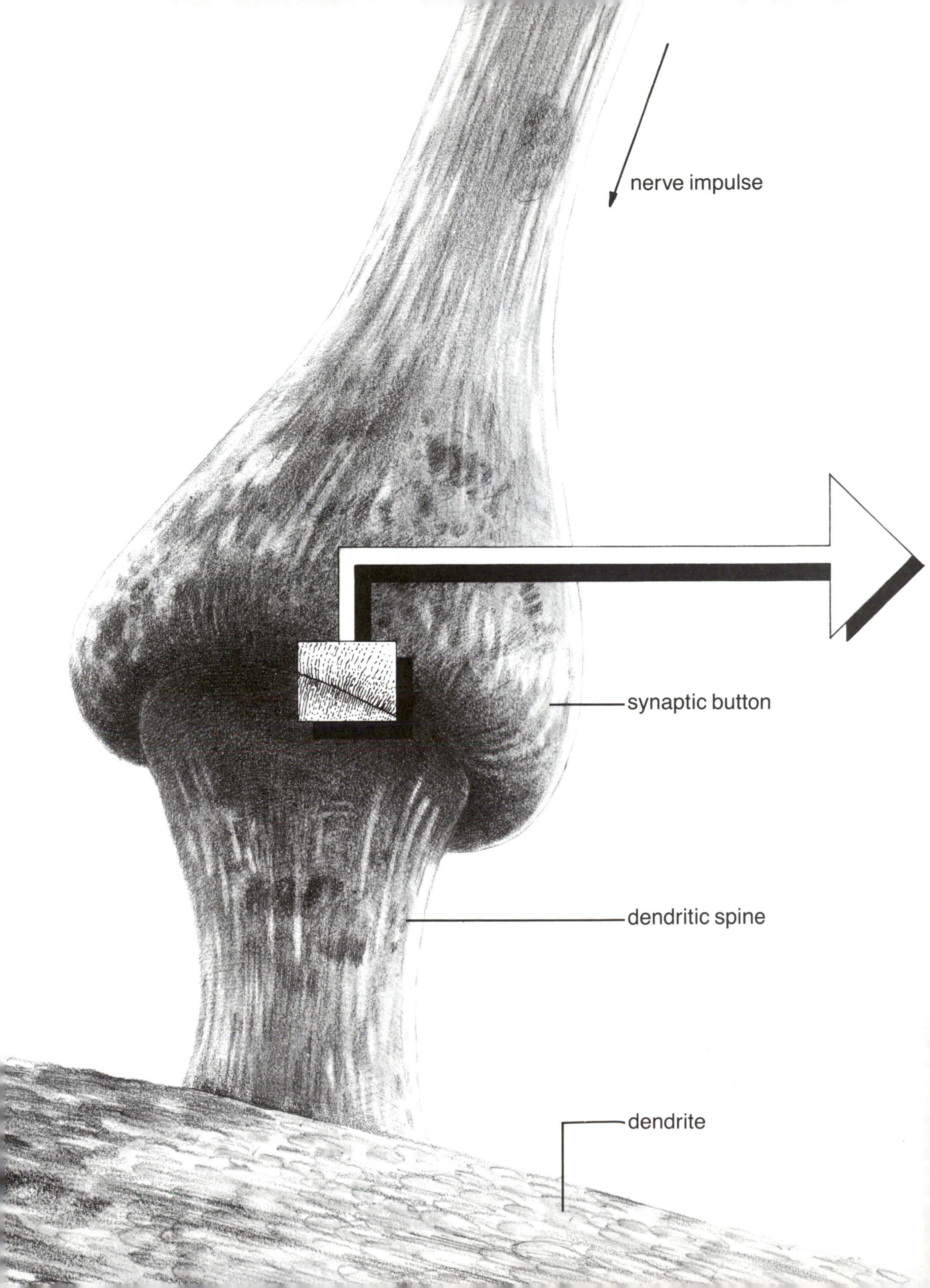
nerve impulse
synaptic button
dendritic spine
dendrite

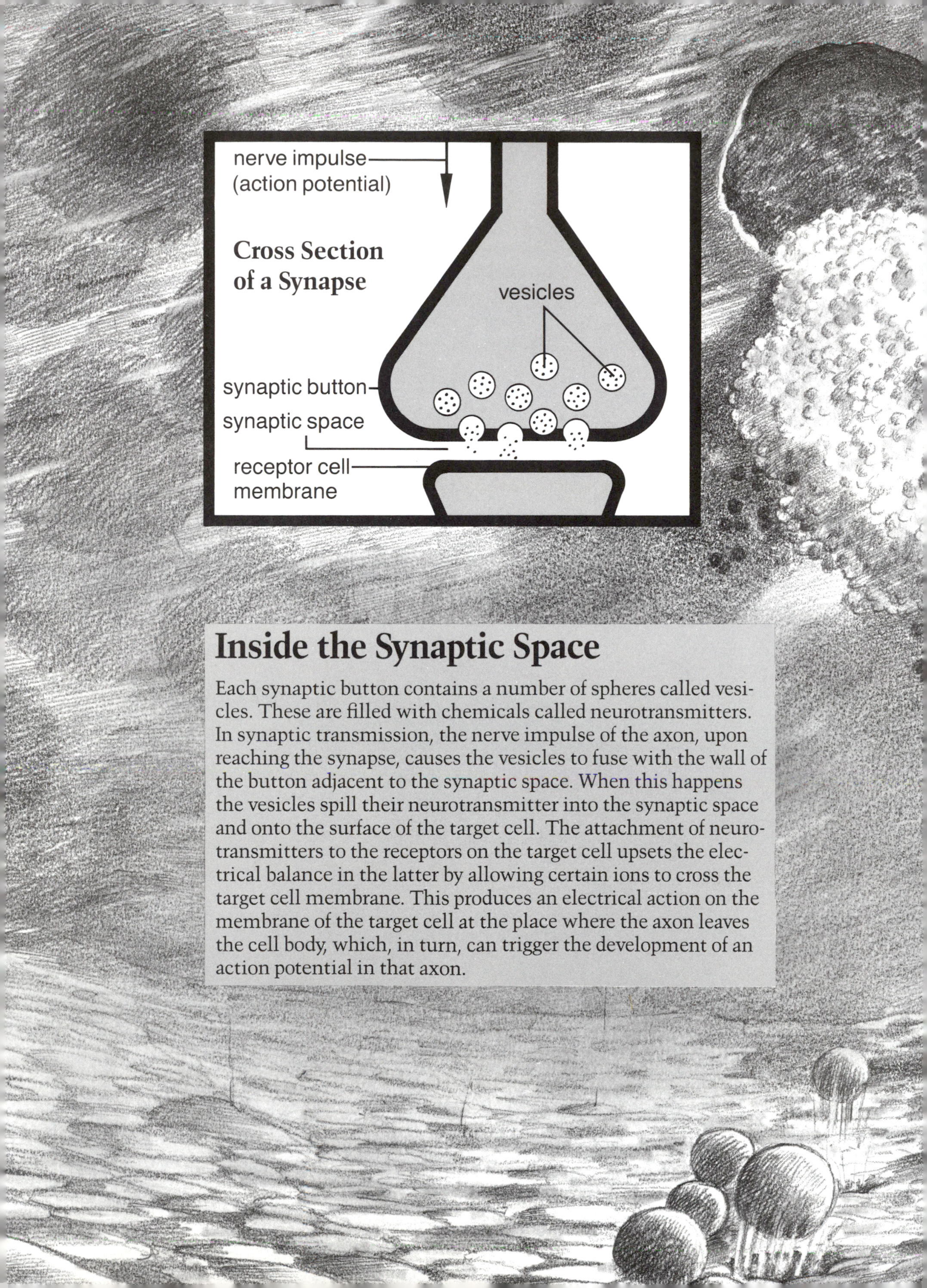

Inside the Synaptic Space

Each synaptic button contains a number of spheres called vesicles. These are filled with chemicals called neurotransmitters. In synaptic transmission, the nerve impulse of the axon, upon reaching the synapse, causes the vesicles to fuse with the wall of the button adjacent to the synaptic space. When this happens the vesicles spill their neurotransmitter into the synaptic space and onto the surface of the target cell. The attachment of neurotransmitters to the receptors on the target cell upsets the electrical balance in the latter by allowing certain ions to cross the target cell membrane. This produces an electrical action on the membrane of the target cell at the place where the axon leaves the cell body, which, in turn, can trigger the development of an action potential in that axon.

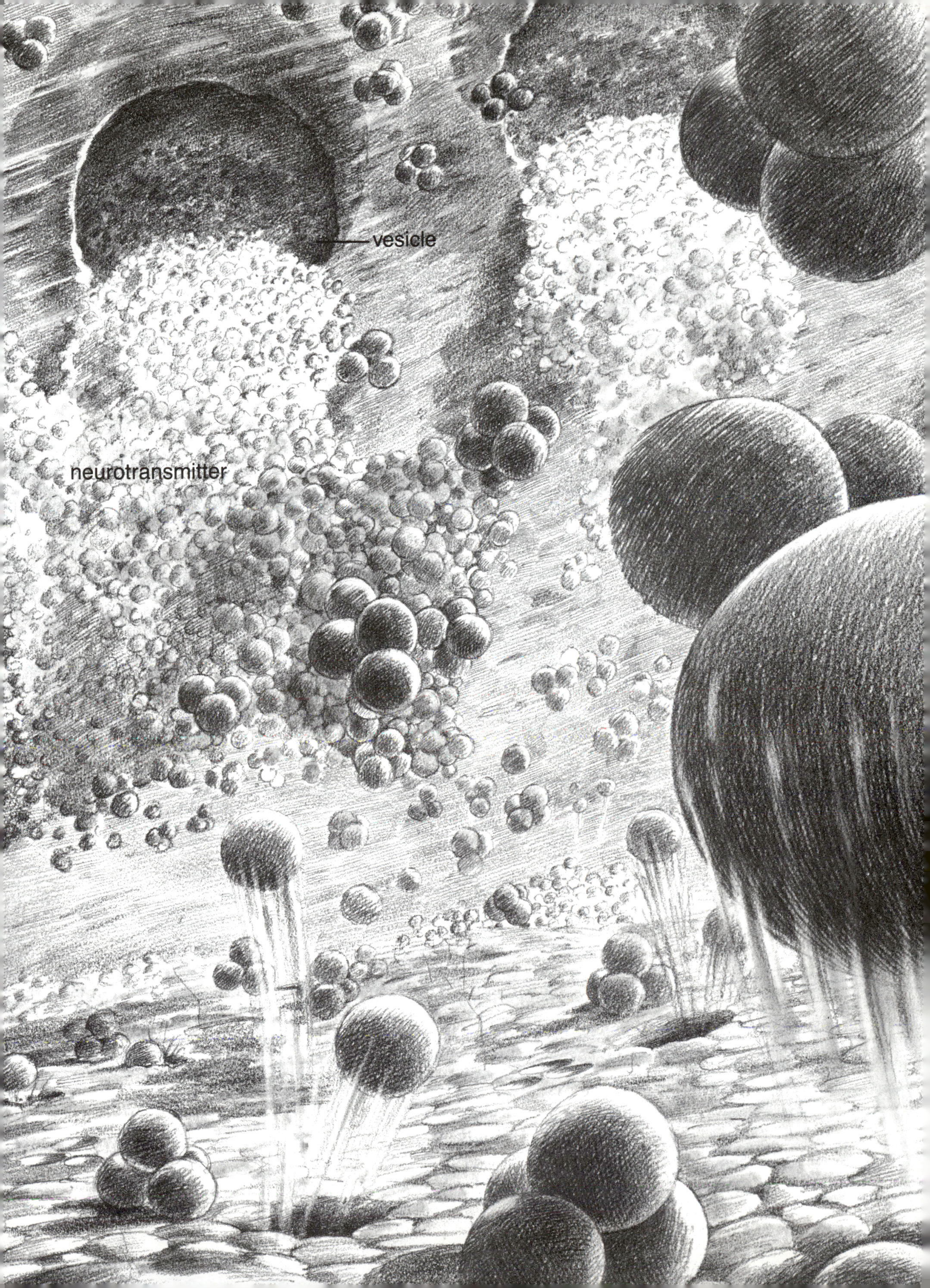
vesicle
neurotransmitter

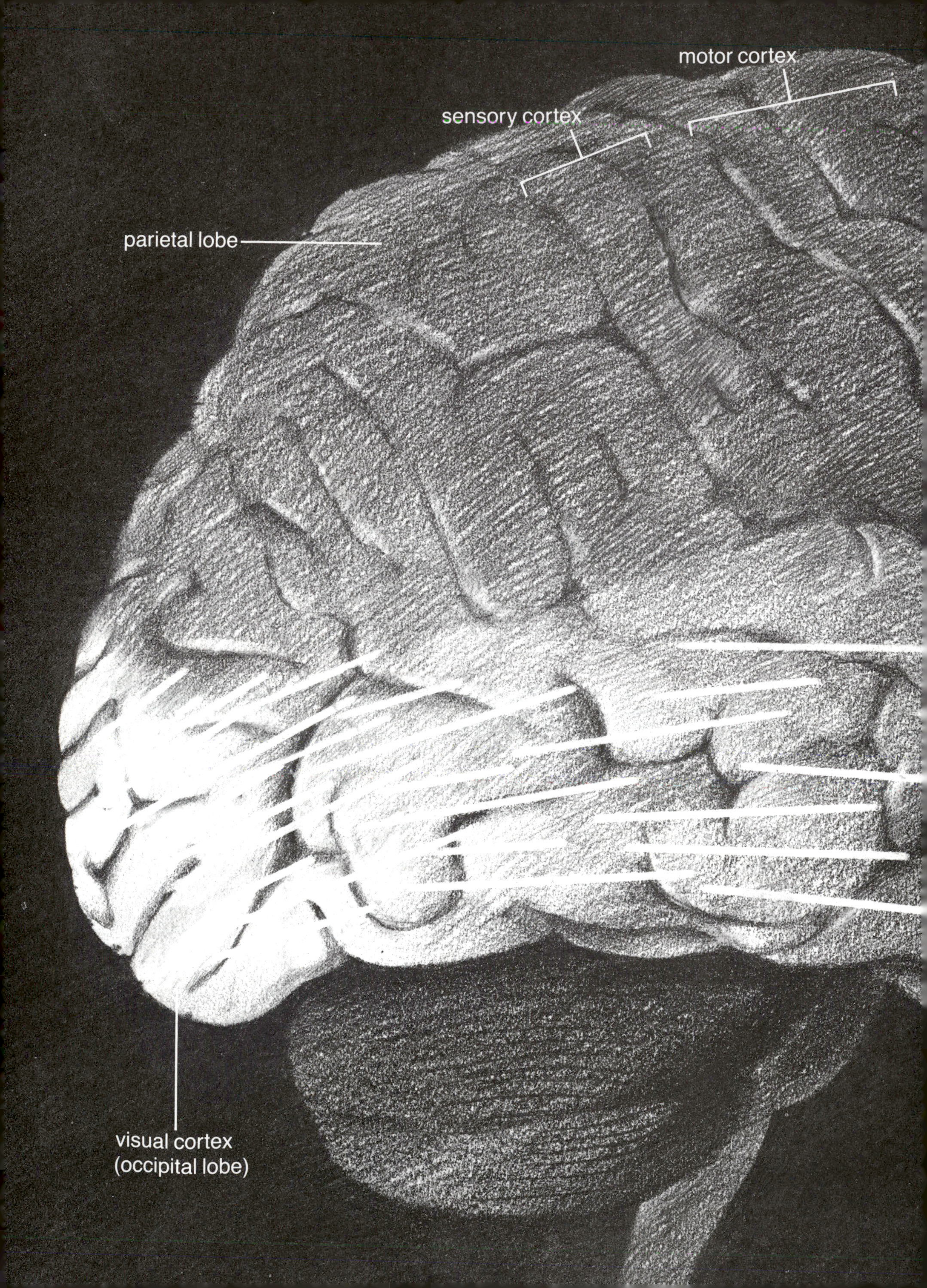

motor cortex
sensory cortex
parietal lobe
visual cortex
(occipital lobe)

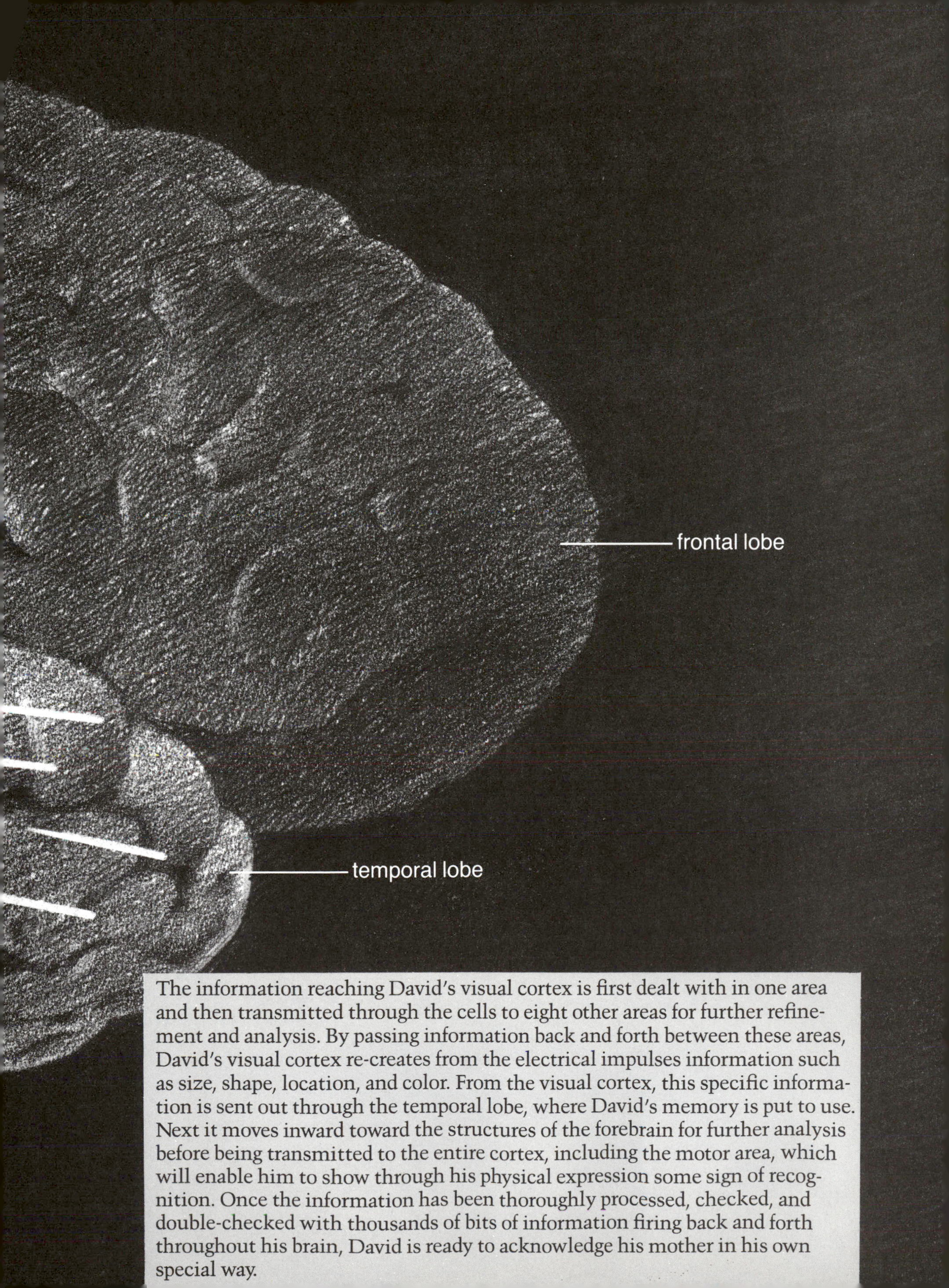

The information reaching David's visual cortex is first dealt with in one area and then transmitted through the cells to eight other areas for further refinement and analysis. By passing information back and forth between these areas, David's visual cortex re-creates from the electrical impulses information such as size, shape, location, and color. From the visual cortex, this specific information is sent out through the temporal lobe, where David's memory is put to use. Next it moves inward toward the structures of the forebrain for further analysis before being transmitted to the entire cortex, including the motor area, which will enable him to show through his physical expression some sign of recognition. Once the information has been thoroughly processed, checked, and double-checked with thousands of bits of information firing back and forth throughout his brain, David is ready to acknowledge his mother in his own special way.

MAMA!

This greatly simplified tour of an incredibly complex chain of events has taken at least several thousand times longer to read than the fraction of a second in which the actual event occurred.

PART II

THE BRAIN, THE MIND, AND THE WORLD THEY CREATE AND REMEMBER

5

Memory: The Changing Brain

THE ABILITY OF THE HUMAN MIND to learn — to store and recall information — is the most remarkable phenomenon in the biological universe. Everything that makes us human — language, thought, knowledge, culture — is the result of this extraordinary capability.

Memories are stored among the neurons of the brain in some kind of relatively permanent form as physical traces, which we call memory traces. If only we knew the code, we could read the entire lifetime of experiences and knowledge from these traces in the brain. This is perhaps the greatest challenge in neuroscience — to understand how the brain stores memories.

A dramatic increase in our understanding of brain memory systems came some years ago from one patient, H.M., who underwent brain surgery to correct a serious epileptic seizure condition. The surgery was successful in treating his epilepsy, but it had side effects so severe that this particular type of brain surgery has not been performed again.

Suppose you are an expert in the study of human brain function and behavior — a psychologist or psychiatrist — and you have been brought in as a consultant to study H.M. You know that he has had brain surgery and you want to see if his mind and his mental abilities are damaged in any way. You also know that he had his surgery several years before and that it cured his epileptic condition. Let us imagine that you are brought into his room in an outpatient clinic and introduced to him. Your interview might go something like this:

You: Good morning, Mr. H.M., it's a pleasure to meet you. My name is Dr. X.

H.M.: Good morning, Dr. X.

You: How are you feeling today?

H.M.: Just fine, thanks.

You chat for a few minutes to put H.M. at his ease. Then:

You: I wonder if you would mind answering a few questions?

H.M.: Not at all. I enjoy being interviewed. It almost seems as if I do it for a living. [He has, in fact, served as a paid subject of study for many years.]

You: Who was the president of the United States during World War II?

H.M.: Franklin Roosevelt and then Harry Truman.

You: Do you remember what President Truman did when a national rail strike was about to occur?

H.M.: I think he nationalized the railway system. Either that or he threatened to.

The questions and answers go on in this vein. You also give H.M. some simple problems to solve and a brief intelligence test. The results are all completely normal. In fact, H.M. is clearly of above-average intelligence and seems in full possession of his mental abilities. You begin to wonder if perhaps you are seeing the wrong patient.

As you continue your interview with H.M. someone knocks at the door to inform you of an urgent telephone call in an office down the hall. You excuse yourself from H.M. and take the call. It is a colleague who has a number of questions about one of your current patients. The telephone conversation lasts about ten minutes. You then return to H.M.'s room.

You: Sorry, Mr. H.M., that my call took so long.

H.M.: Excuse me, do I know you? I don't seem to remember meeting you before.

Suddenly, the awful magnitude of H.M.'s loss hits you. H.M. cannot remember his own experience. You question him further and find that he has no memory at all of having met and talked with you, no memory of the questions you asked or the problems you had him solve. He is quite willing to repeat the process and does so. He performs just as well as before.

As a result of his brain surgery H.M. has permanently lost the ability to learn new things, particularly to remember his own experiences. This loss dates back only to about the time of surgery, actually to a period beginning a few months before the surgery. His earlier memories, his life experiences before the surgery, are intact and normal. H.M.'s remarkable and tragic memory impairment was discovered by the neuropsychologist Brenda Milner a few weeks after his operation.

Surprisingly, there are other aspects of H.M.'s current memory that are not impaired. He has a normal short-term memory. He can remember a new telephone number for a short time as well as you can. However, if you are asked to memorize the number, you can do so by repeating it to yourself over and over and perhaps developing a trick association—a mnemonic device—to remind you of it. H.M. cannot. He is very good at developing trick associations to help him remember things. But the trouble is that it works only so long as he can keep repeating it to himself. As soon as he is distracted from this, he forgets the whole thing—the number and the trick association. It never gets stored in long-term permanent memory.

H.M. also has a normal memory for motor skills. He can learn a complex motor skill like playing tennis about as well as most other people can. But imagine being H.M.'s tennis instructor! You would have to reintroduce yourself each lesson. Suppose H.M. had learned a slice serve from you and did not know how to do it or what it was called before his operation. Each lesson you would have to tell him all over again what a slice serve is. But you would not have to teach him the motions, how to do it. He will learn the actual skill—his slice serve will improve with practice and remain with him as well as it does for other pupils. He just can't remember what it is called or anything you said about it, or, for that matter, who you are. From his point of view, you are a stranger in a new setting each time he takes his tennis lesson.

It is difficult to imagine what it would be like to live forever in the present. H.M. once expressed it poignantly in an interview: "Right now, I'm wondering. Have I done or said anything amiss? You see, at this moment everything looks clear to me, but what happened just before? That's what worries me. It's like waking from a dream. I just don't remember."

The surgery that was done on H.M. removed a part of the brain called the hippocampus (the word comes from the Latin for "sea horse";

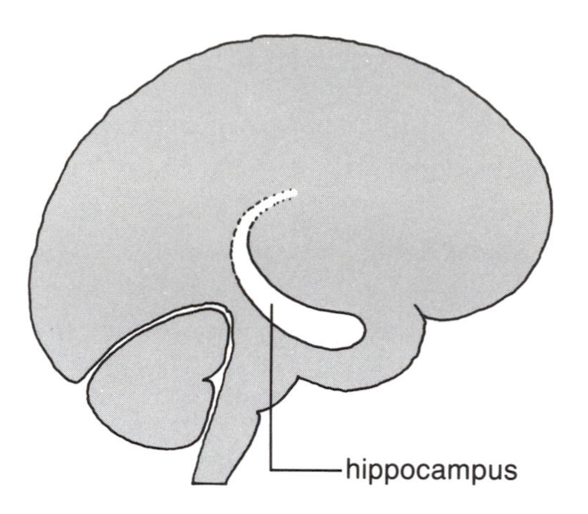

the shape of the human hippocampus resembles the shape of a sea horse). Like almost all other structures in the brain, there is a hippocampus on each side—one in each temporal lobe. Removal of just one hippocampus, either one, does not seem to cause much impairment of memory ability. However, in H.M.'s case, both were removed.

The hippocampus is a part of the limbic brain, the ancient brain system that formed the highest region of the brain in primitive vertebrates like the crocodile. In mammals the cerebral cortex expanded and surrounded the hippocampus and eventually came to dominate it enormously in terms of relative size. In a rat, the hippocampus is almost as large as the cerebral cortex, but in monkeys and humans the cerebral cortex is very much larger. Nonetheless, the hippocampus plays critically important roles in learning and memory in all mammals, including humans, as demonstrated by the case of H.M.

You may recall that H.M.'s memories for his life experiences before surgery were intact. Removing the hippocampus did not abolish those memories; instead, it prevented him from storing new memories after the operation. The hippocampus is not where experiential memories are stored, but it plays a critical role in placing new memories in storage. We think, with much uncertainty, that such experiential memory may be stored in certain regions of the cerebral cortex.

Current research by Mortimer Mishkin, a neuroscientist at the National Institute of Mental Health in Bethesda, Maryland, offers promise of helping to understand the way the brain stores memories. He uses monkeys trained in visual memory tasks as his subjects. The monkey visual system is essentially identical to the human visual system, and visual memory seems similar in monkey and human, although the human visual memory capacity is very likely much greater.

The monkeys are trained on a simple short-term visual-recognition memory task. A tray is first presented with a single small block or toy object covering a food well that contains a peanut. The monkey reaches out and displaces the object and gets the peanut. The tray is removed so the monkey cannot see it, and another tray is presented with the old object and a new object, each covering a food well. But only the well under the new object has a peanut. The monkey must remember and recognize the *old* object to choose the *new* object and get the peanut. On the next trial, entirely different objects are used. The monkeys must learn the principle of always selecting the new object and, of course, remember which was the old object. They learn this task very rapidly; indeed, it is one to which the naturally inquisitive monkey is predisposed.

In discussing the "sensory brain" in chapter 2, we considered how lines, shapes, and objects are coded in the visual cortex. Some neurons respond to the simpler features like edge and orientation and others seem to code more complicated features like shape and color, particularly in secondary visual areas. These areas in turn send converging information about the features of the object being seen to neurons in visual area TE, the area of the temporal lobe where the "monkey hand cell" was found.

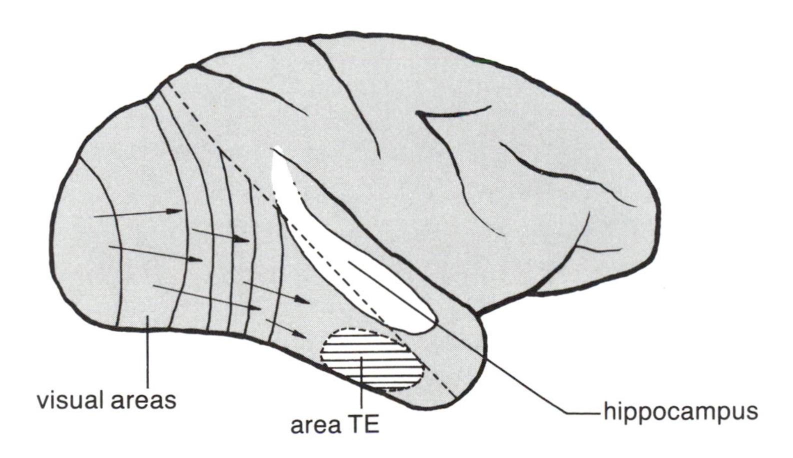

Movement of Visual Information in the Monkey Brain

Mishkin finds that when area TE is removed from both sides of the brain, the monkey loses the ability to perform the visual-recognition task, to remember which was the object he saw before. He still remembers the principle — choose the new object — he just can't remember which was the old object. The same monkeys can still see the difference between objects; they can be trained to choose one and not the other. It appears that their perceptions of objects are still normal. What has been lost is the ability to remember, even briefly, what they have seen.

Destruction of portions of the limbic system including the hippocampus on both sides also markedly impairs the monkey's ability to remember objects he has seen. This brings us back to H.M. Mishkin thinks that there is a circuit from primary visual cortex to visual area TE to hippocampus and back again. It may be that in monkeys, like humans, the hippocampus and other regions of the limbic system play a critical role in placing the memory of the visual object into storage in the cerebral cortex in area TE.

The structure or architecture of memory in the brain is almost completely unknown, one of the great unsolved mysteries, although we have tantalizing hints from current research. Memory must have a permanent structural basis in the brain. Some years ago, Karl Lashley, a pioneering scientist in this field, adopted the term *engram* from the Greek word for "trace" to mean the structural trace of memory in the brain.

How is it that nerve cells store memories? Well-learned memories are permanent, so the cellular storage must involve permanent changes in nerve cells. We do not yet know the nature of these storage processes, but we do know that they must involve chemical changes. If the storage involves growth or structural change in synapses, proteins must be made. Indeed, drugs that prevent the synthesis of proteins in the brain also block the formation of new permanent memory in animals. However, these drugs also have severe side effects, since proteins must be made for many normal functions of the body.

H.M. has a very special kind of amnesia — he can't remember new experiences. Much more common is amnesia for old memories, as in the many stories about people with head injuries who can't remember their past life. It is a very real phenomenon and can be exceedingly serious. A treatment that is sometimes used for certain forms of depression is electroconvulsive shock (ECS), where brain seizures are induced by an electric current applied to the head. ECS causes a temporary amnesia from which the patient eventually recovers fully, except for memories of events just prior to the ECS treatment, as Larry Squire at the University of California at San Diego has shown.

The same kind of temporary amnesia can be produced in rats by ECS treatment, as first discovered by Carl Duncan in 1949. James McGaugh at the University of California at Irvine has studied this intriguing phenomenon for many years. The animal is trained in a simple task—suppose it is placed on a low platform and if it steps off onto a grid floor it gets a paw shock. After one such experience, if the rat is placed back on the platform it will not step off—it has learned in one trial that stepping off has very unpleasant consequences, and it remembers this for many days.

The ECS treatment is then given to different animals at different times after their first experience with the platform. If the ECS is given immediately after the learning experience, the animal loses its memory for the experience. When placed on the platform later, it immediately steps down. However, if an hour elapses between the time of the first experience with the platform and the ECS treatment, the animal later has a perfectly clear memory of its first experience and will not step down again.

Clearly, there is something very different about new memories and old ones. This led McGaugh and others to develop the notion of memory consolidation. Newly formed memories are rather fragile and easily disrupted, whereas older memories are almost impervious to anything short of brain damage. It is as though new memories require some time to wear in or consolidate the permanent memory traces. Perhaps this time is necessary for neurons to make proteins that may be involved in permanent memory storage.

Some years ago it was suggested that memories themselves are actually coded in complex protein molecules, memory molecules. This led to the notion that memory might be transferred from one person to another—protein memory molecules could be extracted from a "donor" brain and injected into another brain to transfer the memories of the donor brain. In a television movie a female scientist-spy was injected with a brain extract from a murdered colleague and relived his memories to solve the case.

Some rather fanciful studies were done on a very primitive flatworm, the planaria, suggesting that memories might be transferred. The "trained" planaria were fed to untrained planaria (these little creatures have an unpleasant tendency to be cannibals) and they were said to have acquired the "memory." There was even a very poorly done experiment with rats suggesting this might occur. The rats fated to be donors were trained to perform a simple task, approaching a food cup. They were killed and their brains ground up and injected into untrained rats. In the report, it was claimed that the untrained rats now remembered the task. This experimental result could not be repeated, even by the person who first reported it. In the end, the planaria studies could not be repeated either. There is no evidence for memory transfer from one brain to another. Memories are not stored as protein molecules. However, the possibility of such memory transfer led *Time* magazine to its infamous final solution for what to do with old college professors: grind up their brains as food for college students.

It may surprise you to learn that all humans have virtually perfect photographic memories. Unfortunately, photographic memory lasts for only about one-tenth of a second. When a visual scene or an array of letters or numbers is flashed before your eyes for just an instant, much of the information in the scene is held accurately in memory very briefly, for a fraction of a second. Most of it is then forgotten. This very short-term photographic memory is called iconic, from the Greek word for image.

Interestingly, most young children have persisting iconic or photographic memory. However, it is lost at the time they begin to learn to read. Anthropologists report that persisting iconic memory is much more common among adults in preliterate cultures, societies where people do not learn to read and write. Learning to read may somehow interfere with photographic memory ability.

When you look up a new telephone number you can remember it just long enough to dial. If you do not repeat it to yourself, it will usually be forgotten in a few seconds. This is often called short-term or immediate memory. Roughly speaking, it is what you hold in immediate awareness at any given moment. It has a surprisingly limited storage capacity for new information—only about seven items can be held in short-term memory. When you have to remember an unfamiliar area code as well as a new telephone number, your short-term memory is taxed beyond its limit. Immediate awareness, of course, includes more than just new items of information—also part of it are sensory experience of the world about you, ideas and thoughts, and well-established memories. It is the ability to hold new information in immediate awareness that is so extremely limited.

If you practice a new telephone number—repeat it over and over—you will manage to remember it more or less permanently. The number of items or "bits" of information stored in the brain of a well-educated adult is very high, at least in the millions. Consider the size of your vocabulary. Each word contains several bits of information. More than that, consider all the faces you have seen in your life. If you see them again, you will recognize a very large number of them. The visual memory ability to remember and recognize faces may be unique to people and perhaps other primates as well. Primates are animals with visual abilities well suited to living in social groups.

It seems that at least some visual information, like that about faces and scenes, may pass directly into permanent memory. In a study of this possibility, a class of students was shown more than two thousand slides, one after the other, each for two seconds. The slides were "home slides"—pictures of people and scenes. The next day, the students were shown all the slides again, but this time two slides were shown each time. Each one they had seen the day before was paired with a new slide they had never seen. For each pair, all the students had to do was to indicate which slide they had already seen. They scored an amazing 90 percent correct.

We think now that the brain systems that store these different aspects of memory may be different. The patient H.M. lost the ability to store permanently new information about his experiences after his operation, but the iconic and short-term aspects of his memory system are normal, as are his old permanent memories from before his operation.

Recent and very preliminary work by neurosurgeons suggests that short-term memory may be stored in a rather small region of the cerebral cortex in the left hemisphere. In one series of studies, the brain was exposed and the patients were maintained on local anesthetics so they could communicate with the surgeons. Using a small electrode, an electrical stimulus was applied to various places on the exposed cerebral cortex to search for abnormal brain tissue (the reason for the surgery). The electrical stimulus temporarily inactivated the cortical tissue directly under the electrode. A rather small and localized region was found where stimulation temporarily inactivated short-term memory ability in the patients.

Understanding of language—permanent memories of the meanings of words—is localized in a region of cerebral cortex on the left hemisphere. Recent studies of neurosurgical patients who are bilingual suggest that the memories for words in the two languages may have somewhat different locations within the language area of the left hemisphere.

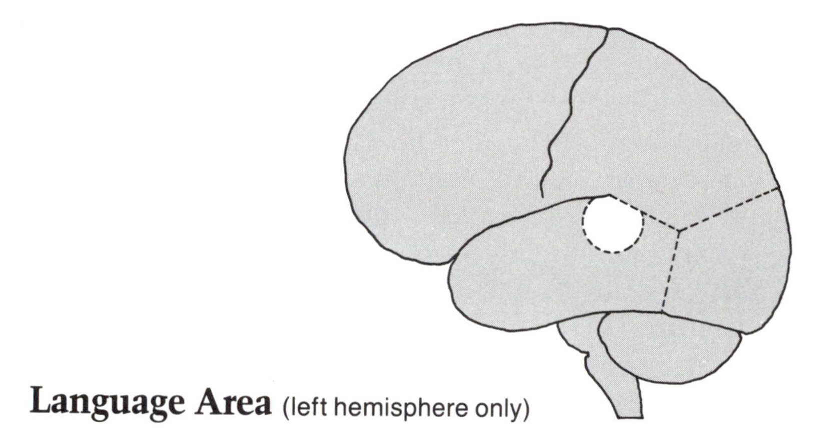

Language Area (left hemisphere only)

In neurosurgical studies done over a period of years by Wilder Penfield at the Montreal Neurological Institute, electrical stimulation of places on the cerebral cortex of the temporal lobe actually seemed to activate very specific memories. One patient was suddenly a child again, experiencing vividly a particular childhood memory. Although this result is very dramatic indeed, the patients being studied were suffering from brain abnormalities that caused severe epilepsy, so we cannot be certain that specific memories could be evoked in this way in normal people.

All these neurosurgical studies at least suggest that certain aspects of permanent memory may be stored in specific places in the brain. But we know so little yet.

Some scientists think that certain kinds of memories are not stored in particular places in the brain but instead are somehow "distributed" widely over the brain, particularly the cerebral cortex. Karl Pribram at Stanford University uses the analogy of a hologram—a three-dimensional image created by lasers. If a part of the screen that contains the hologram image is cut away, the corresponding part of the picture is not cut away, as it would be in an ordinary photograph. Instead, the entire hologram image is still present on the remaining screen, but it is fuzzier. The more screen that is cut away, the fuzzier the picture becomes, but the entire picture is always present. This is, of course, just an analogy — there is no evidence yet that the brain contains holograms.

An increasingly important issue in human affairs is the effect of aging on memory. The process of aging in general is still very poorly understood. The average life expectancy in developed countries such as the United States has been growing progressively longer and is now more than seventy years. However, the maximum life span has not increased but remains about one hundred years. Humans, by the way, are the longest living of all mammals.

The fact that the maximum life span for humans has not increased suggests that there may be built-in aging factors. For a long time it was thought that the problem lay primarily in organs — that the heart, kidneys, and other organs simply got worn out. We know now that this is not the entire answer. Leonard Hayflick of Children's Hospital Medical Center in Oakland, California, has grown cell cultures of normal human body cells taken from people of different ages. Cells from a human embryo double about fifty times before they die. Cells taken from a middle-aged human divide only about twenty times before they die.

This control on cell aging could be in the DNA of the nucleus or in the cell body outside the nucleus. Hayflick exchanged nuclei from human embryo and adult cells and found that the primary control is in the nucleus. Whether the cell bodies were from the embryo or the adult, if the nucleus was from an adult the cell only divided about twenty times. If the nucleus was from the embryo, the cell divided about fifty times.

The mental deterioration that occurs with normal aging has been greatly exaggerated, in large part because of confusion between normal aging and a severe clinical condition of senility called Alzheimer's disease. Donald Hebb is one of the leading scientists in the study of brain mechanisms of learning. In his seventy-fourth year he published an extraordinary paper, "On Watching Myself Act Old." It is a personal account of what it is like to grow old.

Hebb first detected apparent signs of aging when he was forty-seven. He had been reading a scientific paper, and as he read it he thought to himself, "I must make a note of this." He then turned to the next page and saw a penciled note in his own handwriting. It was a terrible shock. He had no recollection at all of having read the paper before. At that time in his career, Hebb was doing extensive research, teaching, and writing, was directing a new laboratory, and was chairman of the Psychology Department of McGill University. He slowed down a little and quit working in the evenings, and his memory soon came back to its "normal, haphazard effectiveness." This is a most important point that is often overlooked—many people, as they reach middle age, take on more and more work and responsibilities. They literally become overloaded. It is not that the memory system begins to fail but rather that although the human memory system has a very large capacity indeed, it is not infinite.

By age seventy-four, Hebb noticed even more changes. His walking and balance had become a little less steady, his vision was poorer, and his forgetfulness had increased a little. He also felt that his effective vocabulary was declining and that his thought patterns tended to repeat themselves, what he termed a "slow inevitable loss of cognitive capacity." However, these losses are not very evident to others. As an editor of the magazine in which the article was published put it: "If Dr. Hebb's faculties continue to deteriorate in the manner he suggests, by the end of the next decade he may be only twice as lucid and eloquent as the rest of us."

Laboratory studies of memory abilities in older people who are not senile indicate that the losses in memory ability with age are not great. Iconic memory is not impaired, although the ability to divide one's attention between two or more sensory inputs may be, as at a cocktail party when listening to the person talking to you and eavesdropping on the next conversation as well. There are no detectable effects of normal aging on short-term memory—the ability to hold information in your immediate awareness. Long-term memory, the ability to store new information permanently, does show a significant decline with aging but not until the sixties or older.

Ten to 15 percent of people over sixty-five suffer from mild to severe symptoms of senility or senile dementia, an unacceptably high percentage of elderly people. Alzheimer's disease has traditionally been defined as severe senility that develops early, before age sixty-five. However, the symptoms are otherwise identical in younger and older people, and senility that develops after sixty-five is now generally included as Alzheimer's disease, or, technically, senile dementia of the Alzheimer's type (SDAT). More than 50 percent of people showing senility can be included in Alzheimer's disease or SDAT—about two million people in the United States.

The symptoms of Alzheimer's disease include marked defects in thought or cognitive processes, memory, language, and perceptual abilities. In some patients the onset is slow and gradual, but in others it can be quite rapid. The first and most obvious symptom is loss of recent long-term memory ability, remembering recent experience and things learned.

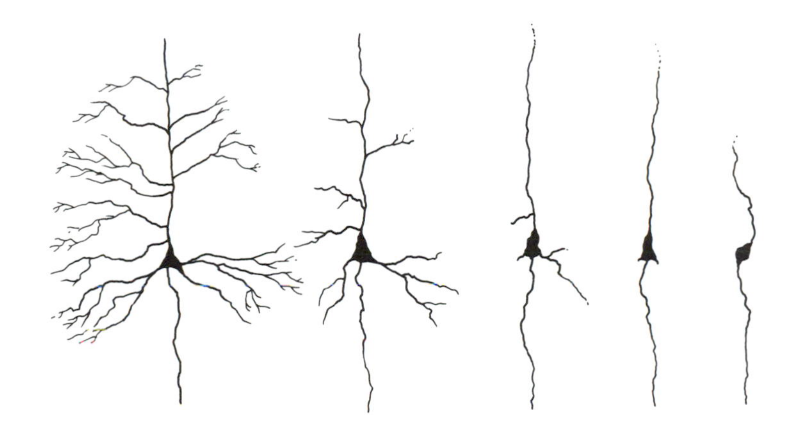

Deterioration of Nerve Cell Dendrites Associated with Alzheimer's Disease

Abnormalities of the brain associated with Alzheimer's disease have been known for some time: senile plaques (clusters of abnormal cell processes surrounding masses of protein), tangles of fibrous material inside nerve cells, a deterioration of nerve cell dendrites, and a loss of neurons. These changes are particularly evident in the hippocampus and certain regions of the cerebral cortex, the brain systems most concerned with complex cognitive processes and memory functions.

A new and exciting story is just developing that may lead to a genuine breakthrough in the prevention and treatment of Alzheimer's disease. It involves a neurotransmitter chemical called acetylcholine (ACh). A few words about ACh: It is the neurotransmitter chemical at the neuromuscular junctions between all motor neurons and skeletal muscle fibers. The nerve impulse causes release of ACh from the motor nerve terminal. ACh activates the muscle fibers, causing muscles to contract. It is then broken down by an enzyme, acetylcholinesterase (AChE), back into its two simple chemical constituents, acetyl and choline. A form of acetyl is present in all cells and choline is present in many normal food substances.

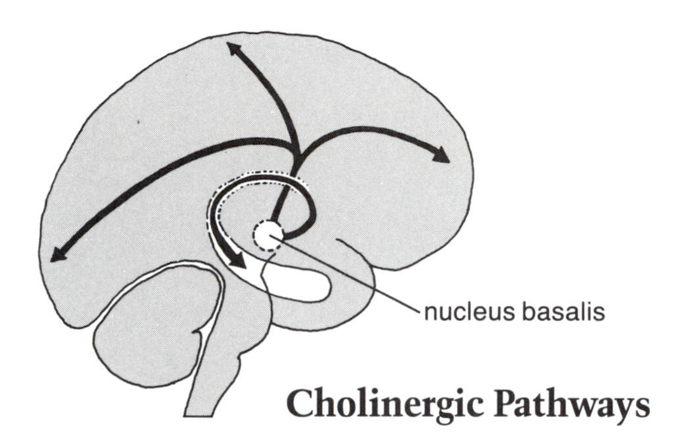

Cholinergic Pathways

There is also a brain system that utilizes ACh as the neurotransmitter. The nerve cells lie near the hypothalamus toward the base of the brain in a structure called the nucleus basalis, which contains the ACh neurons that project to the cerebral cortex and hippocampus. (Neurons that release ACh are called cholinergic neurons.) Some years ago animal studies indicated drugs that increase ACh in the brain seem to improve memory performance.

The recent breakthrough in our understanding of Alzheimer's disease came from work by Joseph T. Coyle and his associates at the Johns Hopkins University School of Medicine. They examined the brains of a number of patients who had died from Alzheimer's disease and found in all cases a massive loss of cells in the nucleus basalis. At the same time, there are much lower levels of chemicals associated with the ACh system in the cerebral cortex and hippocampus in Alzheimer's patients.

It is not yet known whether the loss of ACh neurons is the sole or even major cause of Alzheimer's disease, nor whether there is any causal relationship between such factors as the senile plaques, neuron loss in the cerebral cortex and hippocampus, and the marked loss of ACh neurons in the nucleus basalis. However, now that the striking correlation between the disease and loss of ACh neurons has been established, these questions are more approachable.

ACh drugs can improve memory performance in animals and seem to do the same for normal young adult humans. However, many of the drugs have rather serious side effects; indeed, they are also used as poisons, as in the insecticide malathion, used in California's battle against the Medfly. With the new knowledge of ACh cell loss in Alzheimer's patients, such drugs have become important as therapeutic tools, with some reports of success with patients who are only mildly senile. (A very safe treatment is to eat substances containing choline, such as lecithin or egg yolks.) Feeding choline to very aged mice, who also show deficits in long-term memory ability, is reported to cause improvements in memory performance. But this has not been very successful in humans, at least as a treatment for Alzheimer's disease. Recent reports indicate that a combination of feeding lecithin and administering a drug called physostigmine, which inhibits the breakdown enzyme AChE and results in higher brain levels of ACh, might be of some help to Alzheimer's patients, at least in the early stages.

From a Scientist's Notebook: The Search for Memory in the Brain (Richard F. Thompson)

The brain mechanisms that code and store memories are the primary area of research of one of the authors, Richard F. Thompson. The fundamental problem has been to localize the engrams, or memory traces, in the brain. The cellular basis of memory, how neurons code and store memories, cannot be analyzed until the places and neurons that store memories are found. We were most fortunate recently to discover a tiny region in the brain where certain types of memories may be stored.

We began our search for the engram some years ago, following the lead of ground-breaking scientists such as Ivan Pavlov and Karl Lashley. The strategy we used was to choose a very simple form of learning that is shown equally by humans and animals, particularly mammals, since the basic architecture of the brain is similar from rat and rabbit to human, although much smaller and simpler in the former. We selected eyelid conditioning, a simple situation where a brief sound (a tone) is given, followed by a puff of air to the eye. After a number of such pairings of tone and air puff, the eyelid develops a learned closing response to the tone before the air puff comes. This is a simple adaptive conditioned response learned to try to protect the eye. Rabbits and humans learn the eye-blink response equally well. Indeed, the learned eye-blink response is widely used to study basic properties of simple learning in both animals and humans. Rabbits are docile and cooperative and make good subjects for studies of the brain.

It seemed very likely that the memory for such a simple learned response would not be stored in the highest regions of the brain, regions like the cerebral cortex and hippocampus. Indeed, rabbits with these structures removed can learn the eye-blink response relatively normally. This still leaves a great deal of brain tissue. We were able to rule out some possibilities, such as the motor neurons that control the eye-blink response and the auditory nuclei in the brainstem that relay to the rest of the brain the information that the tone is on.

There was no way to make even a good guess as to where the memory for the learned eye-blink response might be stored. Consequently, we undertook a systematic study to map the activity of nerve cells in all the regions of the brain that might store the memory. To do this we recorded the electrical nerve impulse discharges of nerve cells with a tiny electrode system. This system is fixed to the rabbit's skull while the rabbit is deeply anesthetized. After the animal recovers from this minor operation, it is trained in the eye-blink task and tiny electrodes are inserted in the brain to record nerve cell activity. The brain itself has no feeling of pain or touch, and the animals are quite unaware that electrodes are being inserted into the brain.

These long and laborious brain mapping studies finally paid off. We found a very small region in a part of the cerebellum where the nerve cell discharges increased markedly over the training period. The cerebellum is a large structure of the brain below the forebrain much concerned with movement and had been suggested by some scientists as a possible locus for certain kinds of memory traces having to do with learned movements. The pattern of increased neural activity in this region of the cerebellum actually formed a "model," in time, of the learned eye-blink response to the tone but not of the reflex eye blink to the air puff. Finding this region of neurons that showed learned responses was most encouraging, but it did not establish that this small region was the site of the memory. Some other critical region might simply be relaying neuronal activity about the learned response to this region.

We next made lesions in this region—that is, we destroyed a small amount of brain tissue. After the lesion, the animals completely forgot the task and could never learn it again. However, the eye-blink response to the air puff was still normal—the animals had no trouble making the eye-blink response. Instead, they lost the memory for the *learned* eye-blink response and could never relearn it. We have also found that this region of the cerebellum is essential for another kind of learned response that is widely used in the laboratory—learning to lift the leg to avoid a shock to the paw. It appears to be where the memories may be stored for a whole class of simple learned responses.

The critical locus for destroying the memory of the learned eye-blink response is very small. Tiny chemical lesions that destroy no more than one cubic millimeter of nerve cells destroy the memory. Several other lines of evidence in our current work point to this tiny region of the cerebellum as being where the memory is stored, but we have not yet proved that beyond all possible doubt. However, having identified a critical part of a memory circuit, we can now identify the entire memory circuit, from "ear to eyelid," which we are now in the process of doing. This will allow us to establish with certainty the exact location of the memory trace. We will then be in a position to tackle the most important question of all—how it is that neurons code and store memories in the brain.

At the cellular level there are two fundamental types of information coding or memory. One of these is the genetic code. In higher animals literally millions of bits of information are coded in the DNA of the cell, the genetic memory. This information is vast, and it determines not only whether we will be a mouse or a man but the myriad characteristics that make up each individual.

Over the course of evolution a quite different kind of information coding has developed—the cellular encoding of memory in the brain. This memory code is no less remarkable than the genetic code. As we have seen, a well-educated adult has literally millions of bits of acquired information stored in the brain.

The fundamental difference between the genetic code and the memory code is, of course, that each individual human's memory store is acquired through experience and learning. The uniqueness of each human being is due largely to the memory store, the biological residue of memory from a lifetime of experience. We will someday understand the genetic basis of the ability of the brain to store memory, but we can never know the actual memories that are stored from studying the genes, only from studying the brain.

The ability to learn is an emergent property of cellular tissue. The ability, as such, has a clear genetic base; it is dependent on the structural and functional organization, the architecture, of the brain and on cellular storage processes. It would not be entirely surprising if the genetic material itself plays a role in the process of learning. After all, the actions of nerve cells at synapses can engage the interior of neurons and even act on the DNA itself.

6

The Divided Brain

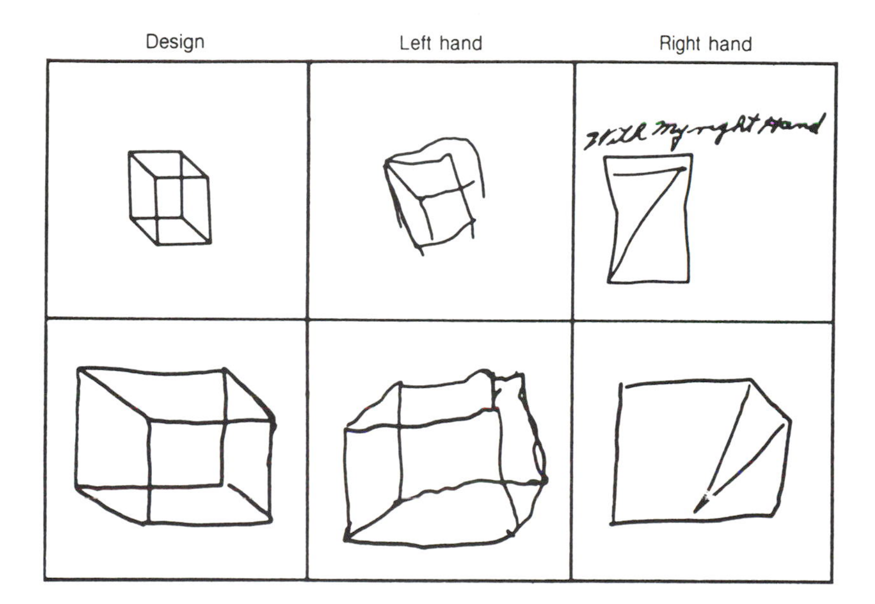

YOU CAN GO FOR A WALK with them, go swimming, eat dinner, even ride in a car they are driving, and not notice anything. You wonder what the radical brain surgery is about. They seem normal: their coordination is fine, their reason is unimpaired, their ability to dance, to walk — to sing, even — is fine. One of them can even play the piano better than the scientist interviewing him. But these people have had a very radical brain operation for epilepsy, splitting their brain, and the results have caused a revolution in the understanding of the most evolved part of the brain.

But it is only under very subtle circumstances that the differences in the hemispheres are seen, only when a special effort is made to discover them. Look at the drawings on the previous page. They are done by the same person, one of the "split-brain" people—patients who have had the connection between the hemispheres severed. Although the surgery has left the patients relatively unimpaired, you can immediately see the difference in these drawings: one is almost indecipherable; another is simply a representation of a cube, poorly drawn, but recognizable nonetheless. How this operation came about and what has happened since is a story that is now about one and a half centuries old, one of the longest continued investigations in the study of the amazing brain.

To go back to the beginning: In 1834 and 1835, Marc Dax, a French physician, began to collect evidence on people who had lost their ability to speak as a result of a brain injury—the condition known as aphasia. In reviewing his cases, Dax noticed that all of them had damage to the left side of their brains. He undertook a search of the literature and of his colleagues' experiences, since Dax was an ordinary doctor (something similar to our general practitioner) and not well versed in the neurology of his day. There were no cases of aphasia that did not involve the left hemisphere, as far as he could find, although some patients, of course, had damage to the right hemisphere as well.

In 1836 Dax went to a medical association meeting in the south of France and presented his carefully documented evidence: there is a strong association between damage to the left hemisphere and the loss of language. The response was underwhelming: Dax received no notice and his contribution seemed to be forgotten. But as with many important findings in science, it was waiting to be rediscovered.

In the nineteenth century the idea began to arise that the brain was not just an amorphous lump of tissue, as the Greeks had thought, but that different areas of the brain might have different functions. Evidence began to mount that this was indeed the case, and it would have been accepted further had the understanding of that evidence not been faulty—for the proponents of "cerebral localization," as it was then called, were also proponents of phrenology.

The doctrines of phrenology contained the idea that bumps on the surface of the head mirrored the differences in the size of the underlying brain tissue, so that the kind of brain a person has would then be "read"

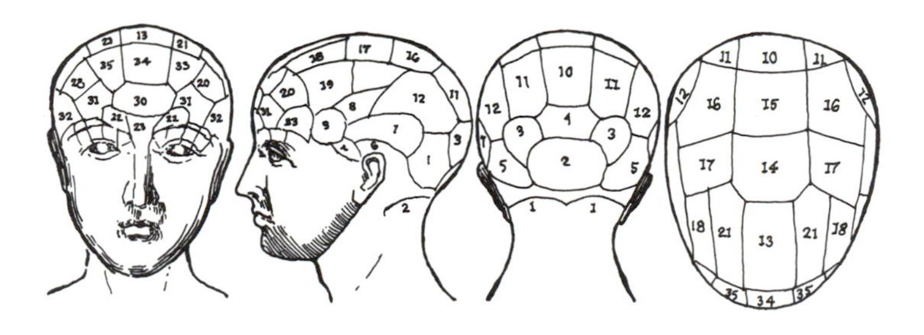

Illustration from an 1829 Broadside Announcing Lectures on Phrenology

on the skull. Different areas of the brain were assigned different functions and representations on the head. This idea had two fatal flaws: it was easily tested, and it was wrong. Thus the idea that the hemispheres were different was buried in the controversy over the obviously inane phrenology, another setback.

The problem began to be settled by the French neurologist Pierre Paul Broca, some twenty-five years after Dax had presented his careful report. As an unwilling participant in the phrenology controversy, Broca had simply begun to examine the brains (after they had died) of people who had known cases of speech loss. He presented eight very carefully studied and documented cases, all of which had speech loss and all of which had damage in a certain portion of the left frontal lobe. This

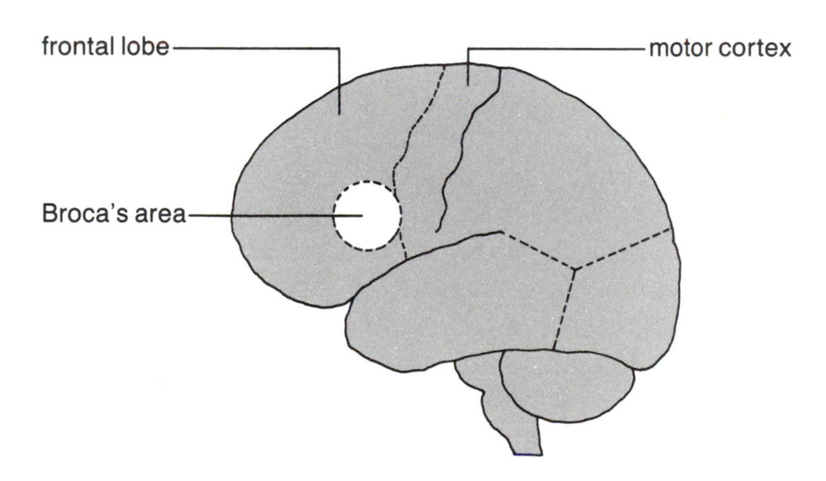

Broca's Area (left hemisphere only)

area became known as Broca's area, one of the parts of the brain involved in speech production. It was the first generally accepted evidence on brain asymmetry, to the chagrin of Dax's son, who mounted a campaign to revive interest in his father's primacy. But the field was established by Broca's careful work, and so for more than a century now, the neurological evidence has come from the study of people whose brains have been damaged by accident or illness and from the surgery performed on them. It is, then, in the work of clinical neurology and neurosurgery that the primary indications of our hemispheric specialization are to be found.

In 1864, after reading Broca, the great British neurologist J. Hughlings Jackson concluded the left hemisphere to be the seat of the "faculty of expression" and noted of a patient with a tumor in the right hemisphere, "She did not know objects, persons, and places." Since Jackson's time, many other neurologists, neurosurgeons, and psychiatrists have confirmed that two different modes of thought seem to be lateralized in the two cerebral hemispheres of human beings.

In thousands of clinical cases studied, damage to the left hemisphere very often interferes with, and can in some cases completely destroy, language ability. An injury to the right hemisphere does *not* destroy language performance in most cases, but it may cause severe disturbance in spatial awareness, in musical ability, in the recognition of other people, and in the awareness of one's own body. For instance, some patients with right-hemisphere damage cannot dress themselves adequately, although their speech and reason remain unimpaired.

In precise neuropsychological studies, Brenda Milner and her associates at the Montreal Neurological Institute have attempted to correlate disorders in specific kinds of tasks with damage to specific areas of the brain. For example, removal of the right temporal lobe severely impairs the person's ability to find his way out of a maze, while left temporal lobe damage of an equal extent produces little loss. These researchers also report that damage to specific parts of the brain results in specific kinds of language disorders: an impairment of verbal memory is associated with the damage in the front of the left temporal lobe; speech impairment seems to result from damage to the rear of the left temporal lobe. On less empirical grounds, the Russian physiologist A. R. Luria writes that mathematical function is also disturbed by lesions of the left side. Milner and her associates find that the recognition of musical pitch seems to be in the province of the right hemisphere. Other laboratories associate a loss of the ability to recognize faces with damage to the rear of the right hemisphere.

The clinical neurological research is intriguing, correlating the different functions of the hemispheres that are impaired by brain damage. Perhaps the most intriguing is the research of Roger W. Sperry of the California Institute of Technology and his associates, notably Joseph Bogen. Sperry's work intrigued the Nobel Prize committee enough to win him the prize in 1981. As we have seen, the two cerebral hemispheres communicate through the corpus callosum, which joins the two sides anatomically. Sperry and his colleagues had for some years experimentally severed the corpus callosum in laboratory animals, which allowed the hemispheres to be trained independently. This research led to the adoption of a radical treatment for severe epilepsy in several patients of Drs. Philip Vogel and Joseph Bogen of the California College of Medicine.

This treatment involved an operation on humans similar to Sperry's experimental surgery on animals — cutting the interconnections between the two cerebral hemispheres, effectively isolating one side from the other. This surgery came to be called split-brain surgery. The hope of the researchers was that when a patient with severe epilepsy had a seizure in one hemisphere, the transmission of the seizure between hemispheres would be lessened, and the seizure would therefore diminish. The surgery worked and in most cases the patients were improved enough to leave the hospital.

In day-to-day living, these "split-brain" people, described at the beginning of this chapter, exhibit almost no abnormality, which is somewhat surprising in view of their radical surgery. However, Sperry and Bogen developed several ingenious and subtle tests that showed that the operation had clearly separated the specialized functions of the two cerebral hemispheres.

If, for instance, the patient held a pencil, hidden from sight, in his right hand, he could verbally describe it, as would be normal. But if the pencil was in his left hand, he could not describe it at all. Recall that the left hand informs the right hemisphere, which possesses only a limited capability for speech. With the corpus callosum severed, the verbal (left) hemisphere is no longer connected to the right hemisphere, which communicates largely with the left hand. If, however, the patient was offered a set of objects—a key, a book, a pencil, and so on—and was asked to select the previously given object with his left hand, he could choose correctly, although he still could not state verbally just what he was doing. This situation closely resembles what might happen if I were secretly asked to perform an action and you were expected to discuss the action, about which you'd been told nothing.

Normally, when we wish to inquire about the knowledge of another person, we allow the verbal apparatus to determine it—that is, we reduce "knowledge" to that which a person can report. The preceding example is a primary indication that doing so can be a fundamental error. *We are aware of more than we can discuss.*

Another experiment tested the lateral specialization of the two hemispheres, using split visual input. The right half of each eye sends its messages to the right hemisphere, the left half to the left hemisphere. In this experiment, the word *heart* was flashed before the patient, with the *he* to the left of the eyes' fixation point, and the *art* to the right. Generally, if a normal person were asked to report this experience, he or she would report having seen *heart.* But the split-brain patients responded differently, depending on which hemisphere was responding. When the patient was asked to name the word just presented, he replied *art,* since this was the portion projected to the left hemisphere, which was answering the question. When, however, the patient was shown two cards—one with the word *he,* the other with the word *art*—and asked to point with the left hand to the word he or she had seen, the left hand pointed to *he.* The simultaneous experience

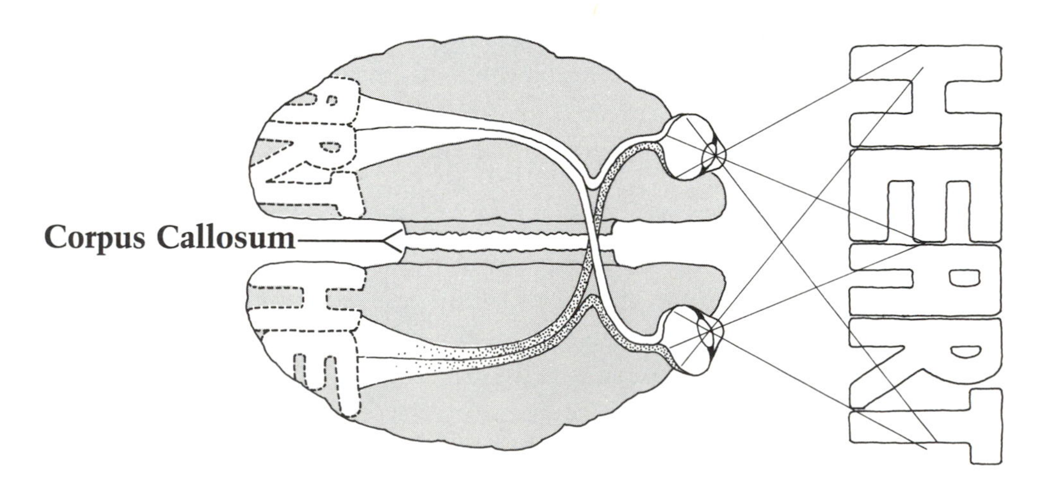

of each hemisphere seemed unique and independent in these patients. The verbal hemisphere gave one answer, the nonverbal hemisphere another.

One of the most dramatic instances of the superior ability of the right hemisphere was captured on film by Roger Sperry and one of his colleagues. The split-brain patient is given a set of blocks, with different divisions of red and white paint on each side. He is asked to assemble the blocks to match certain patterns presented to him. The patient begins well, with his left hand, putting blocks into more and more complex patterns. Then he is asked to begin to assemble them with the right hand.

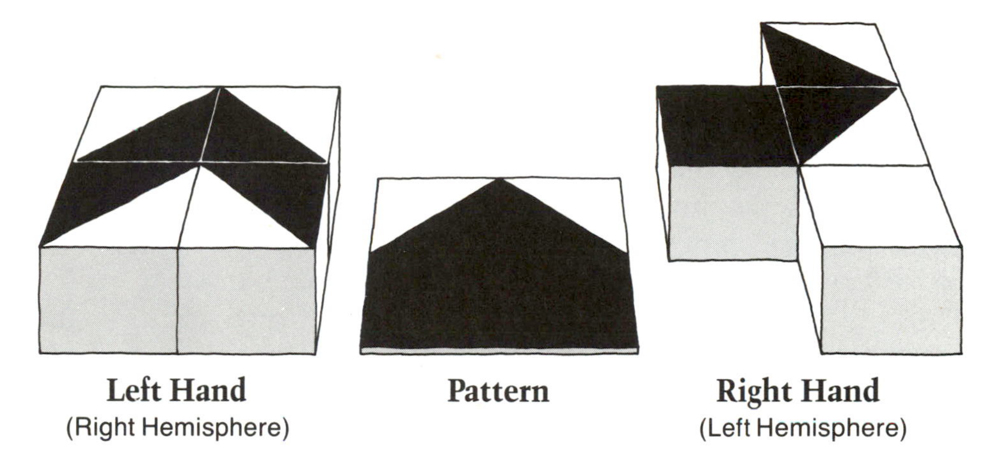

Most people assume that it would be quite easy to do the same patterns with another hand; after all, the person has just seen himself doing it. However, the right hand has great difficulty; even in the simplest of patterns it turns over the blocks seemingly at random. At one point in Sperry's film the patient turns over a block to complete a pattern, but then he continues turning it over to make an incorrect pattern, much to the viewer's dismay. But here is the interesting part: the patient's left side is also dismayed. The left hand appears, furtively, on the side, and attempts to correct the right, only to be chastised by the experimenter.

Joseph Bogen, one of the surgeons who performed the operation, studied the effects on drawing ability after the split-brain operation (the effects are shown in the drawings at the beginning of this chapter). In another study, Bogen asked one of the patients to copy both a cross

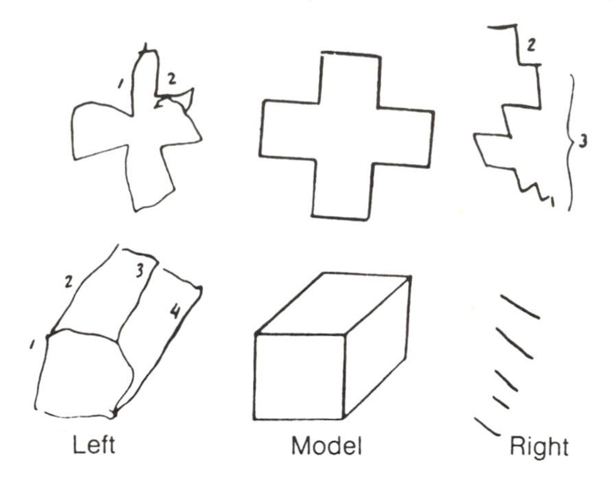

and a cube. The illustration here contains the drawings made by the left and the right hands. The copies by the left hand (the right hemisphere, remember) are fairly good representations of the original figures; the drawing quality of those done by the right hand is slightly better, but the communication of the form is terrible: as with the drawings at the beginning of this chapter, there is little way to know what the original form was. The parts of the whole are simply "listed," in an order, as it were. "Everything we have seen," Sperry writes, "indicates that the surgery has left these people with two separate minds, that is, with two separate spheres of consciousness." And, it should be noted, these "spheres" are quite different: one seems able to express itself in words, the other in drawing and in space. But the question remains, do normal people operate this way, or is the split brain just a unique surgical production?

These spectacular split-brain and lesion studies are not the only evidence for the physiological duality in consciousness. In general, great caution should be exercised in drawing inferences to normal functioning from pathological and surgical cases alone. In dealing with these cases we must remember that we are investigating disturbed, not normal, functioning, and the connection to normal functioning may be a bit tenuous. In cases of brain damage, it is never fully clear that one hemisphere has not taken over a function from the other to an unusual degree because of the injury. In the case of the surgical patients, they are by definition quite atypical, owing to their epilepsy.

So, to complete the story it is necessary to get evidence from normally functioning people, even if that evidence is necessarily more indirect, since we don't go poking inside the brains of such people. In this, we are fortunate that recent research with normal people has confirmed much of the neurosurgical explorations. The evidence comes from many sources — tests of vision, eye movements, reaction time, ear preference, and electrical signs of brain asymmetry.

From a Scientist's Notebook: The Specialization of the Cortex (Robert Ornstein)

Against this background, one of the authors, Robert Ornstein, has tried to develop an easy method to study whether the normal brain actually does make use of the "lateralization" it shows when split. The brain was "built" upward from the brainstem, each layer of this "ramshackle house" having been placed over the previous one. This means that the most interesting parts of the brain — the different parts of the cortex — are right under the skull, and the activity of the brain actually can be recorded by placing sensitive electrodes on the surface of the skull.

This recording is called an electroencephalogram, or EEG, as it is known in the trade. The EEG consists of an examination of the voltage the brain produces, recorded from the surface of the skull; the voltage is very low, on the order of a few millionths of a volt. The EEG is a rather crude measure, something like recording the overall noise that a city produces. If you did try to make such a recording, you might find that there are large amounts of noise in the center of the city from 9:00 to 5:00, more in the outlying areas after dark, and very little noise after midnight. But you would hardly use the overall noise measure to determine anything very subtle, such as how the inhabitants were going to vote in the next election. By the same token, recording brain waves tells us only when one part of the brain is "noisy," or active.

The idea behind the experiments we designed was that if the brain of a normal person activated different hemispheres while thinking, then by recording the EEG from both hemispheres of a normal person working at a cognitive task we might be able to see a sign of the selective activation and suppression of these two hemispheres.

With colleague David Galin, Ornstein tried out these ideas on a trusting medical student assigned to them one summer. He was outfitted with EEG electrodes over the left and right temporal and parietal

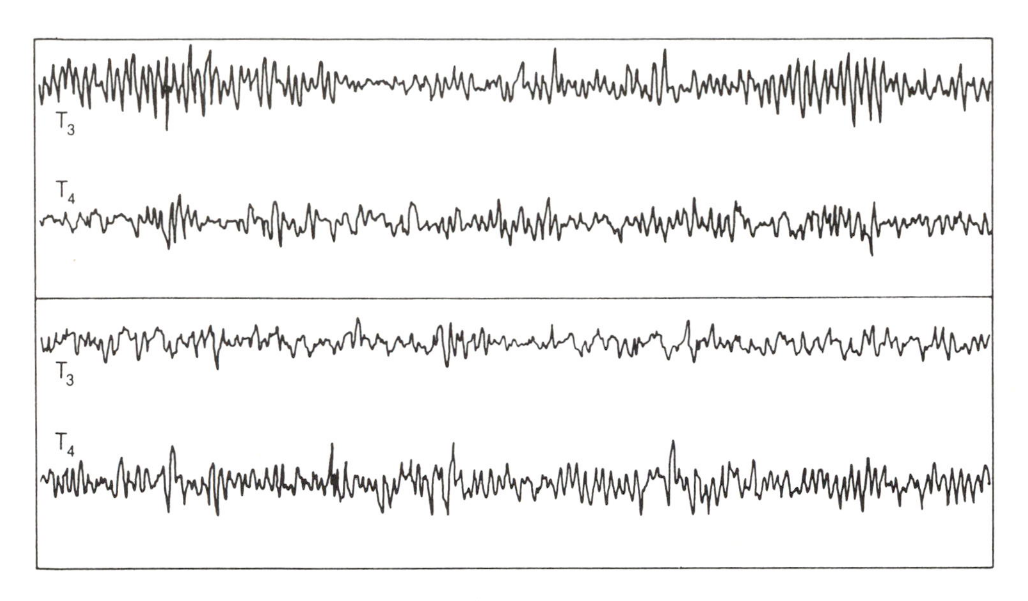

EEGs Recorded During Two Different Cognitive Tasks

areas of the skull. Then he was asked to perform verbal and spatial tasks: writing a letter and arranging a set of colored blocks to match a given pattern.

The findings were immediate and very striking: while writing (presumably a left-hemisphere task) he produced high-amplitude EEG alpha waves (waves at approximately 10 cycles per second) over the right hemisphere and much less amplitude over the left hemisphere. This pattern reversed while he was arranging blocks, with the alpha rhythm dominant over the left hemisphere and less visible over the right hemisphere. The alpha rhythm is generally taken to indicate a diminution of information processing (similar to the noise of a city shutting down) in the area involved.

This seemed to be what we were searching for, a measure of the activity of the two hemispheres in a normal person. The left hemisphere "quieted down" while our student was arranging the blocks; the right hemisphere quieted down while he was writing. We retested him all summer and also recruited other laboratory personnel for testing. With the new subjects we found similar results: their EEGs showed (for each task) that the area of the brain *not* being used was relatively "turned off."

Consider the EEGs (shown opposite) recorded while normal people were engaged in these cognitive activities. Note the appearance of the high-amplitude alpha in the left hemisphere (marked T_3) during the presentation of spatial problem situations, and the greater alpha in the right hemisphere (T_4) during work on verbal and arithmetic tasks.

With these results in mind, we considered the many difficulties other researchers have encountered in attempting to relate EEGs to intelligence, cognition, and consciousness. Fortunately, we had attended to several factors that seem to have been neglected in the past: (1) We recorded each EEG while the subject was engaged in a task, rather than trying to relate a "resting" EEG, or averaged evoked potential, to subsequent performance. (2) We selected cognitive tasks that clinical evidence has shown to depend more on one hemisphere than the other, which, therefore, would be associated with a predictable distribution of brain activity. (3) We selected EEG-electrode placements on anatomical grounds. A wealth of evidence suggests that temporal and parietal areas of the brain should be the most functionally asymmetrical, and the occipital areas the most similar to each other. Unfortunately, occipital EEG recordings have been used most often in the past, probably because they are not as sensitive to extraneous interference from the muscles and eyes.

We recruited ten new people for a formal study of this phenomenon. The people were asked to write a letter, to arrange wooden blocks, to match a design, and also to perform mental versions of these tasks (mental letter writing and mental matching of forms). We analyzed the result in terms of the ratio of total right-hemisphere power to left-hemisphere power in both temporal and parietal pairs. We interpreted higher power in the EEG to mean more idling; a high ratio therefore meant more right-hemisphere idling and indicated a more active involvement of the left hemisphere in a task. Similarly, a low ratio indicated a more active involvement of the right hemisphere. The technical results showed that during block manipulation and mental forms the ratios were 10 to 20 percent lower than during the verbal tasks. This indicates more left-hemisphere involvement in verbal tasks, more right-hemisphere in spatial tasks.

More recent studies using this measure show that the primary factor in hemisphere specialization is *not* the type of information (words and pictures versus sounds and shapes) considered, but how the brain processes the information. A recent study compared subjects' brain activity while reading two types of written material: technical passages and folk tales. There was no change in the level of activity in the left hemisphere, but the right hemisphere was more activated while the subject was reading the stories than while he or she was reading the technical material. Technical material is almost exclusively logical. Stories, on the other hand, are simultaneous: many things happen at once; the sense of a story emerges through a combination of style, plot, and evoked images and feelings. Thus, it appears that language *in the form of stories* can stimulate activity of the right hemisphere.

In another experiment, brain activity was recorded while subjects mentally rotated sets of boxes. This operation normally involved the right hemisphere. When they were asked to do the task analytically, by counting the boxes, subjects by and large "switched over" to their left hemisphere. So, people can use their hemispheres differently in problem solving.

An important extension of this research is found in the work of Richard Davidson and his colleagues at the State University of New York at Purchase in the past few years. If a person is asked to relive an intensely emotional experience, the hemisphere involved depends on the nature of that emotion. There is a recent and very popular myth that has it that everything good must involve the right hemisphere. In fact, it is the *left* hemisphere that seems to be involved in happy and pleasant emotional experiences, and the right that is involved in negative feelings such as anger. This startling finding is made even more so by a further study by Davidson: in recording ten-month-old babies, the same result was found: the right hemisphere is more involved in emotions such as anger, the left in happiness.

There is as yet no satisfactory explanation for this, but the speculation runs as follows: the left hemisphere may be involved in fine motor control, the right hemisphere in the control of large motor movements such as running and throwing. It might be that it was useful in our evolutionary history to have the control of large movements placed closely in the brain to the focus of negative feelings, so that if something had to be done, such as running or hitting, it could be done quite soon. Be that as it may, this is a new and important finding and one that will further extend our knowledge of how functions are placed in the brain.

These split- and whole-brain studies have led to a new conception of human knowledge, consciousness, and intelligence. All knowledge cannot be expressed in words, yet our education is based almost exclusively on the written or spoken word. One reason it is difficult to expand our ideas of education and intelligence may be that as yet we have no standard way of assessing the nonverbal portion of intelligence.

The two ways of knowing are not competitive but are complementary. Without a comprehensive perspective, our ability to analyze may be as useless to us as it was to the right hand of the split-brain patient. Similarly, an intuitive insight is lost unless we have a way to express it. Many people whom we consider "unintelligent" or "retarded" may in fact possess a different kind of intelligence and may be quite valuable to society. The neurologist Norman Geschwind has put the dilemma this way:

> One must remember that practically all of us have a significant number of special learning disabilities. For example, I am grossly unmusical and cannot carry a tune. We happen to live in a society in which the child who has trouble learning to read is in difficulty. Yet we have all seen some dyslexic children who draw much better than controls, i.e., who have either superior visual-perceptual or visual-motor skills. My suspicion would be that in an illiterate society such a child would be in little difficulty and might, in fact, do better because of his superior visual-perceptual talents, while many of us who function well here might do poorly in a society in which a quite different array of talents was needed to be successful. As the demands of society change, will we acquire a new group of minimally brain-damaged? [Norman Geschwind, "Language and the Brain." *Scientific American* 226(4):76–83.]

7

The Individual Brain

IT SEEMED a straightforward job. All one of the authors (Robert Ornstein) was trying to do was to measure the different sizes of the cortex in several brains, to get an idea of the size of the different areas of the brain underneath the skull. He didn't realize how much there was to be discovered by actually looking at real brains. What was most striking was this: like most people, the author's idea of what the brain is came from anatomical drawings, the models he had seen, and the sample brains he had dissected. But as he worked in the laboratory day after day, he began to realize that each brain was distinct. One had characteristic bulges here, one there, one a large occipital lobe, another a small temporal lobe. In fact, people's brains are as different as their faces.

Faces have, of course, some regularity: the eyes are above the nose, the nose is above the mouth, both are above the jaw. But within this regularity there are wide variations: some faces have large noses, some have small eyes. So it is with the brain. The specific features in the brain are different in different individuals. How the brain develops, how it grows and changes within a person's lifetime, and even how it changes within a day is an area just beginning to be explored. We may not think for long of "the brain" as exactly the same in all people. Some of the recent discoveries cited here will make that idea impossible.

You will recall from chapter 1 that humans are born extraordinarily immature, and that the human brain develops largely in the outside world. So environmental conditions play a greater role in the brain development of humans than in that of any other primate.

It is commonly thought that at birth the neurons begin to make connections and that these connections increase as we age and acquire experience. However, the opposite appears to be the case: there may be many *more* connections and nerve cells in the brain of an infant than in an elderly adult. Development seems to be more a matter of "pruning" those original connections than of making any new ones. Consider this about infant babbling: in the first weeks of life, a baby utters almost every sound of every known language. Later on, the infant loses the ability to make sounds that are not in the language he has learned to speak. There may be an enormous potential of sound patterns available to us at birth, but we learn only a few of them. The brain may be "set up" to do many different things, such as to learn the thousands of languages available to humans, only a few of which we actually do learn.

However, the growth of the brain depends on an adequate early environment. Severe malnutrition may cause inadequate brain development, a smaller brain than normal, and severe mental retardation. In a long series of experiments, rats deprived of normal food in infancy show distortions in brain structure, and even shrinkage of certain brain structures. Cells from "deprived" ones look shriveled compared to normal cells. The brain, somewhat like a muscle, then, can grow in response to certain experience—the neurons themselves become larger.

Some of the most revolutionary evidence has come from a series of studies over the last twenty years from work initiated by Mark Rosenzweig and continued by Marion Diamond at the University of California at Berkeley. They study rat brains so that they can control the genetic background. Rats have a fairly short gestation period—twenty-one days—and they have, most beneficially for these purposes, a smooth cerebral cortex. The dog brain is folded, the cat brain is folded, but the rat brain has yet to fold, and that's one of its beauties for making chemical and anatomical measurements—the smooth-surface cerebral cortex allows one to deal with uniform pieces of tissues.

All the animals are in "standard colony" conditions for preliminary measures, which means that there are three rats to a small cage and water and food are provided. Besides standard colonies, the experiment involves ones with environments enriched with "toys," or objects to play with, and ones with impoverished conditions, in which there is little stimulation and movement is restricted by cage size. The enriched condition in the postnatal rat consists of twelve rats living together with toys. Every day the experimenters change the objects from a standard pool. If they don't change them, the animals become bored, just as we all do when we've sat too long receiving the same type of stimulus. In the impoverished environment, the rat lives by himself and has no toys; he can see, smell, and hear the other rats but does not play with them.

Typically, Diamond selects three brother rats; one goes into enriched, one into standard colony, and one into impoverished environment. Even in young adults a year old, the enriched environment will cause an increase in the actual *weight* of the brain—about 10 percent in most cases. At first most scientists didn't believe the results, but the evidence has now convinced virtually everyone.

Although their work was revolutionary enough, Diamond and her colleagues were curious to see if they could produce the same result in the brains of very old rats. They put four very old rats in with eight of the young to see whether the stimulating effect of associating young rats with old would result in measurable changes in the brain. It turned out that the old rats enjoyed living with the young more than the young enjoyed living with the old. The brain growth of the rats confirmed this result. *Each* old rat's brain grew by 10 percent while living with the young rats. The young rats' brains did not grow at all while living with the old. Why didn't the young rats' brains grow while the old rats' brains did? A clue may be found in the different responses of the old and young to the experimental situation. Each day when one of the experimenters went to change their toys, the old rats would come to see what toys were available, while the young remained sleeping in the back. So it appears, says Diamond, that there is some sort of hierarchy when the young live with the old; the old dominate. Marion Diamond often jokes that this is why old professors continue to be excited and live to a hundred—because they are dealing with young people, who are essentially like the young rats. And everybody knows that it is the young students who are sleeping in the back of the lecture hall.

An analysis of the brain growth showed that the specific changes in the brain took place in the dendrites of each nerve cell, which thickened with stimulating experiences. It is as if the forest of nerve cells became literally enriched, and the density of the branches increased; this is what produced a bigger brain.

Not only specific experiences can affect brain growth. Such conditions as increasing negative ionization of the air (the kind of "charged" air found on mountaintops, near waterfalls, or at the sea), when introduced by a negative ion generator to Diamond's rat colonies, produced the same changes in brain growth. So, not only do friends and stimulating experiences get into your head and brain, but so might the fresh air of mountaintops, waterfalls, and other places where the ion concentration (both positive and negative) is greater. The ions can also change the chemical composition of the neurotransmitters, and can elevate or suppress mood, something almost everyone knows who has noted the exhilaration of the mountains, or the depression with a Santa Ana wind.

Whether diet can change brain growth has not yet been studied, but foods have been found to influence the chemical environment of the brain. The brain is precious tissue and is specially protected from the outside world: the skull is a solid shield, and there are also internal barriers to protect the brain. A special network of cells called the blood brain barrier prevents poisons in the bloodstream from reaching the brain. Also, although the brain is only 2 percent of body weight, it uses 20 percent of the oxygen. Because of such protections, the inner workings of the brain have been thought to be almost completely isolated from the state of the body or from happenings in the external world.

This is wrong. Diet affects us even before birth. Harvey Anderson of the University of Toronto has found that the diets fed to nursing mother rats can significantly affect the food preferences of their young —carbohydrate-eating mothers, for instance, produce carbohydrate-eating offspring. In another study with more ominous implications, Bernard Weiss of Rochester University has found that toxic additives fed to pregnant rats are registered in the fetus as quickly as ninety minutes later.

Some recent experiments by Richard Wurtman of the Massachusetts Institute of Technology have found striking short-term changes in brain chemistry associated with diet. In one early study, eating proteins was found to increase serotonin production in the brain. Eating substances rich in choline dramatically increased the neurotransmitter ACh throughout the brain, particularly in the brainstem and cerebral cortex. Choline is present in lecithin (which is sold as a food supplement), in egg yolk, and in lesser amounts in fish, cereals, and legumes, so eggs may be more of a "brain food" than fish. Whether specific diets can help in learning and memory is far from proven, but it would be foolish to ignore the research possibility. Other foods also have specific effects on the workings of the brain. The amino acid tryptophan, when ingested, increases the production of neurotransmitter serotonin, thought to be involved in the regulation of sleep and waking. These fascinating, if early, findings open an important area of inquiry. The brain is subject to important short-term changes with diet: food preferences, toxic effects on behavior, alertness, sleepiness, all may be influenced by the food we eat, as it changes the chemical operation of our brain.

As the brain can change in response to long-term environmental conditions, it also can rearrange its organization to compensate for accidents and changes of demands. Although in most people language is in the left hemisphere, people with left-hemisphere damage can be trained to produce language using their right hemisphere, although this flexibility decreases with age. The right hemisphere takes on language functions in young children who have suffered severe damage to the left hemisphere. In the deaf, areas of the temporal cortex normally used for the processing of speech sounds are used instead for processing visual information.

A striking instance of this capacity occurs when a person learns a second language. There are intriguing reports that when the second language is learned, brain organization can sometimes change: in one case, the first language migrates from the left hemisphere to the right. In others, the second language may occupy only the right hemisphere, or it may be represented in both.

The brain seems flexible and adaptable, and this capacity to change may underlie the differences between the brains of different people: here we will consider three important groups—right- and left-handers, people of different races, and men and women. The brain of the left-hander is different from the right-hander's. There are three different kinds of brain organization of left-handers. One group seems to have the hemispheres divided as do right-handers, one group is *reversed,* and a third has language and spatial abilities in both hemispheres. Whether these differences translate into deficiencies is equivocal: some studies find both language and spatial defects in left-handers, some do not. Left-handers (and, surprisingly, their relatives) *do* recover from brain damage better than do right-handers, which indicates that their language abilities are most dispersed in the brain.

What is less equivocal is the cultural bias against things of the left.

The word *gauche* (meaning awkward) is the French word for left. The word *sinister* is the same as the Latin word for left.

Whether brain differences manifest themselves as personality or intellectual traits is unknown, but the existence of strong brain differences is certain, and they may result in differences in body regulation. Left-handers, for instance, have higher rates of autoimmune diseases than right-handers.

There is a clear genetic difference in left- and right-handers, as, of course, between the sexes. But what about race? Are there racial differences in the brain? It is difficult to be completely positive about this, but racial differences seem quite unlikely. Race itself is a dubious concept—the genes that specify skin color and eye folds do not appear to be strongly related to genes that specify personality and intellectual abilities. And, people get their genes from their parents, not from a so-called racial pool. Furthermore, the evidence on "intelligence" (however it is defined) and "race" is so complex that no real judgment can be made. If there *are* any racial differences, they are easily overcome by training. Not so with sex.

Here we come to the most controversial point. There do exist profound differences in men's and women's brains, differences that are present, often, before birth. There have been in recent years many bits of evidence that document differences in behavior and aptitude: girls are more verbally fluent than boys, have better fine motor control, and are less aggressive. Boys have better control of the large muscles, are more sensitive to movement, and are more aggressive. What is new is that these behavioral differences have physical expression in the brain. Boys show earlier right-hemisphere development than girls. Sandra Witelson of McMaster University asked boys and girls three to thirteen years old to match objects held in their hands to visually presented shapes. At five, boys showed a superiority in performing the task with objects held in the left hand (compared to the right). Girls do better than boys on left-hemisphere tasks in grade school years.

In addition to the differences in hemispheric maturation, the hemispheres in males are more specialized than those in females. The representation of analytic and sequential thinking is more clearly present in the left hemisphere of males than in females, and spatial abilities are more lateralized in the right hemisphere of males than females. Thus, damage to the left hemisphere interferes with verbal abilities more in males than in females and damage to the right hemisphere interferes with spatial abilities more in males than in females.

Only recently has an important bit of evidence on the male and female brains been discovered. While examining the corpus callosums of several brains, Christine De Lacoste and her colleagues, in work beginning at Berkeley, found that they could begin to identify them as male and female by sight, just as Robert Ornstein could begin to identify individual brains on meeting them again and again. The corpus callosums of the men as opposed to women studied so far are as different ("dimorphic" in the jargon) as are men's and women's arms: an observer can easily group them into the different sexes by sight alone. The women's corpus callosums are larger than the men's, and they are larger toward the back of the brain. This is the area of the brain involved in the transmission of information about movements in space and about visual space. It is just the area of the corpus callosum that one might expect to be different, given that spatial abilities like throwing seem to be less lateralized in females, involving both sides of the brain rather than just one. This difference appears as early as twenty-six weeks in utero; that is, it is an inborn difference in the major system of brain communication.

Whether we will find more differences is another question, but now that we know that the brain becomes physically different with different kinds of experiences, different conditions of the air, different learning situations, different foods eaten, different handednesses, and different sexes, at least some of the pieces of the puzzle of why humans are so different from one another may be found. There are probably more differences in human brains than in any other animal partly because the human brain does most of its developing in the outside world. We are not as interchangeable as are other animals, and the root of this is in our brain. Neurophysiologists can produce an accurate atlas of the cat's brain, so similar are the structures of one cat's brain to another. Imagine trying to do this with people—even the corpus callosum is different in men and women. This main difference surely is one of the things that makes us such a stimulating and irritating bunch to each other.

8

The Brain: Our Health Maintenance Organization

We are in a race with ourselves. It is in part a race that takes place in our brain.

Human life and the human brain are different, as far as we know, from other species. This difference is the wellspring of our creativity as well as the soul of many of the continual human problems. It results, in no small part, from the spectacular growth of the human brain in the past few million years.

All species have evolved to live within the confines of their original habitat—animals have made radical physical adaptations such as flying and hibernation to suit their specific environments. Humans, however, are different: we have gone outside our original confines of East Africa and the Fertile Crescent. We now live all over the earth, in crowded cities, in freezing weather, in skyscrapers, and even, for brief periods, outside the planet itself.

We have not changed biologically in the past twenty thousand or so years, but the changes we have made in our environment are dramatic. We have constructed a new world for ourselves—cities, let alone airplanes, didn't exist twenty thousand years ago. The challenges we face are different from those of any other species. Our environment is changing faster and faster. We humans must adapt to each change we make in the world, and our ability to make those changes is increasing.

So the problem is that our ability to create always leaps ahead of our ability to adapt, and we are forever locked into a cycle of adapting to unprecedented situations.

If there are too many changes in our lives, such as the death of a spouse, a new and demanding job, or a move to a new city, our ability to adapt may be taxed to the limit and we may become ill. Part of the problem is the ramshackle nature of our brain: sometimes it is capable of handling the challenges of the environment, sometimes it is not. It is quite fashionable now to assert that we are thus living on borrowed time, far beyond our ability to adapt. This is partly true, but as we learn more about the brain, we find that it does have an ability to govern our eating, our weight, and our health in some of the most trying circumstances.

First we'll look at some of the problems, then the happy ending.

You are watching a murder movie on television. You tense, then relax; your heart begins to pound, your mouth gets dry, your stomach may turn, and your hands get clammy. This happens several times just as the killer comes into view, then recedes. Finally it is over, he is captured, and you return to everyday life. What you went through is the "emergency reaction"; it is an old, innate biological reaction to prepare you for the unexpected. It involves heart rate increases as well as changes in the liver and spleen, respiration, pupils, and muscles. You are ready to react. When you take a new job or show up for a date, you react similarly. People activate this reaction far more in contemporary society than ever before. The reaction is prehistoric, useful for emergencies, but not when we are constantly facing changes. It is as out of date as goose flesh—the body's attempt to keep warm by raising up a nonexistent layer of fur so that it may trap the air.

In modern society, however, the number of changes that we experience is far, far greater than we were "designed" for. No one was designed to see fifteen thousand murders on television before the age of fifteen, or to experience constant urban noise and changing conditions. No one was designed to go from the stagecoach to the space shuttle within a lifetime. So, we are rooted in our evolutionary past, yet the creativity in our newly grown brain has made us literally "reach for the stars." The result is that we too often break in the middle, as we move into new and unexpected situations in life. Indeed, it has been found that the more changes we experience (stressful life events such as buying a house, moving, divorce—or marriage!) the more likely we are to become ill. Although not everyone who experiences a series of these changes gets ill, this important finding tells us something about our biology and society: our current environment is, for many people, beyond their biological limits.

Too much life stress can cause heart disease. The striking rise of heart disease in this century is not entirely the result of changes in diet, exercise, cholesterol, and smoking. These factors account for only half the occurrences of heart disease. Sir William Osler, a physician practicing at the turn of the century, noted that his typical patient with coronary disease was "not the delicate, neurotic person . . . but the robust, the vigorous in mind and body, the keen and ambitious man . . . whose engine is always at full speed ahead."

People who have heart attacks often appear healthy, but they have an exaggerated biological reaction to the stresses of their lives. This has been called Type A behavior. Type A, coronary-prone individuals are described as fast-paced, impatient, and irritable; they are deeply involved in their work, and deny failure, fatigue, and illness. They try to get more and more done in less and less time. They do not care very much about relationships with their co-workers, but are very anxious about the opinions of their supervisors.

Type A's are twice as likely to develop heart disease as Type B people, who may be as successful as Type A's, but tend to be calmer, better organized, less "time urgent," concerned more with quality than quantity of work, and less prone to frustration.

But how does stress cause disease, anyway? We now know a fair amount about the mechanisms of the general physiological reactions to stress, but we are just beginning to understand the more specific ones. The emergency reaction is under the control of the brain. It stimulates the heart to beat faster and directs peripheral blood vessels to clamp down, thus increasing blood pressure. The principal neurotransmitters that accomplish this are epinephrine and norepinephrine. In addition, the sympathetic nervous system stimulates the secretion of a great deal of epinephrine and some norepinephrine, which become doubly assured of reaching the target organs.

The specific physiological reaction that may damage the heart, therefore, is the Type A's extreme responsiveness to situations. Type A's show greater emergency reaction to challenge than do Type B's. They shift more often and more extremely into the emergency reaction of high heart rate and blood pressure (as well as the other associated changes) and back again; the constantly increasing and decreasing blood volume can directly weaken the arterial walls. The blood clots faster during the emergency reaction, which increases the rate of atherosclerosis, the formation of deposits on the walls of the arteries. This prevents blood from reaching the heart muscle and thereby starts the process that leads to heart attack. In addition to this decrease in the heart's blood supply, epinephrine and norepinephrine may precipitate a heart attack, because, in addition to making the heart beat faster, they make the heart beat irregularly (this is the condition known as cardiac arrhythmia). The brain's control systems are intimately involved in coronary events.

Another important recent discovery concerns the immune system, which is the body's defense against disease and other toxins; as Jonas Salk once pointed out, it has functions similar to the brain. It is a highly complex system involving many components. Many researchers now believe, first, that the brain regulates the immune system and, second, that the state of the immune system is more important in the development of diseases than exposure to toxins or viral or bacterial disease entities. Some viruses, such as herpes simplex, are always present in the body, but they only become active when something goes wrong with the immune system. Cells that can become cancerous constantly circulate in the body, but in healthy people they are routinely eliminated by the immune system. These "mutant" cells can only take root when some factor, either genetic or environmental, has suppressed the functioning of the immune system.

The immune system, therefore, is thought by some to hold the key to curing or preventing cancer and a wide variety of other illnesses, perhaps including schizophrenia.

Some of the most exciting research on the brain today involves the effect of psychological processes, especially stress, on immune system functioning. Both life events and some personality characteristics affect susceptibility to and recovery from disease. For instance, one recent study showed that how women cope with breast cancer has a greater impact on recovery than the size of the tumor or the type of treatment.

New techniques have allowed researchers to directly measure indicators of immune system functioning, and they are now beginning to link emotions with some changes in immune system reactions. For example, ten weeks after the death of a spouse, bereaved subjects had a tenfold decrease in one kind of immune system responsivity.

We are just beginning to understand how mental states such as grief affect the immune system. There are many parts to the immune system, and each may bear its own relation to psychological processes, but the reaction is controlled by brain processes, some of which are modifiable.

That we are under stress is a phenomenon too obvious to be denied. The reigning conception among physicians and scientists is that our health problems will get worse and worse as we move further and further beyond our biological inheritance and the world moves further and further beyond our control. But to prematurely assert the brain isn't quite up to coping with the stress of modern life does not do justice to its amazing capacity to regulate our health.

What should be wondered about is how we manage to stay as healthy as we do in our complex environment. Most people who are under stress do not get sick, most people who smoke do not get lung cancer, most people who grieve do not die quickly, most people who move and live entirely new lives remain healthy. Our body temperature remains constant, our heart makes billions of beats on time, our glands receive the correct chemical messengers, and millions of other regulatory processes go on, almost automatically. The brain is designed primarily to run the body and to keep it healthy. The countless extensions of the brain's sensory systems, the internal nervous, chemical, and regulatory systems, all serve to keep us out of trouble. The brain is our largest organ of secretion — it produces the most chemicals of any organ in the body — and it is the organ of health, our own internal health maintenance organization.

Some of the most recent discoveries in brain sciences have enabled us to get a glimpse of how elaborate our innate healing network is, and how much we might be able to accomplish were we able to develop drugs and procedures that allowed this innate network to flourish.

In one important study at the University of California at San Francisco, under the direction of Jon Levine, a large number of dental patients were given one of several drugs before dental work. Some were given painkilling drugs, as usual, but others were given a placebo — an inert substance that is believed by the patient to produce genuine physiological effects. Both groups reported little or no pain while in treatment. This finding, so far, is similar to many throughout the world—that inert substances can, if they are believed genuine, influence the body.

This "placebo effect" has often been maligned in medicine, as if nothing "real" is accomplished. This is similar to the experience of Robert Esdaile, the first person to demonstrate hypnosis to the Royal Society, the esteemed British scientific organization. Esdaile, in front of an assembled society committee, sawed off the gangrenous leg of a patient, on stage, without anesthesia. But his treatment was not accepted. Members of the Royal Society alleged that Esdaile had merely hired a "hardened rogue" to appear. So, too, is the placebo ignored, and it is often considered trivial by those interested in "hard" medicine.

But Levine's experiment was different. After administering the placebo to some patients, Levine did something else, something quite innovative. He gave half the patients a dose of naloxone, which, as we have seen, is a drug that blocks the effects of endorphins by filling the receptor sites so that the endorphins cannot operate. If the placebo were merely foolery, then naloxone should have no effect. But if the placebo activates our endorphins, naloxone would have an effect. The results were astonishing to many in neurochemistry: those patients who were given naloxone did *not* produce the placebo effect; they found the dentistry painful. This means that the placebo effect, in this experiment anyway, may have involved the production of endorphins by the dental patients, owing to their belief that they were going to be relieved from pain.

So the brain may be able to relieve pain by producing chemicals on demand that block transmission of the pain signals. Endorphin production has been reported to influence weight, memory, schizophrenic-like symptoms, and many other bodily functions. Even more tantalizing, the immune system has endorphin receptors.

So the brain seems to possess capacities for healing and self-repair beyond the dreams of researchers only a few years ago. It seems to be able to regulate our health far beyond anything that could be done consciously: Norman Cousins reported that laughter helped him overcome a mysterious disease; Augustin De La Peña has hypothesized that the excessively bored brain may be responsible for some cancers; Alan Frey has found that emotional tears may contain substances the body needs to expel; and there are numerous new ideas about how "mental" health and physical health are similar. Here we focus on an important and continuous program of the brain, maintaining a proper weight.

The need to trust innate mechanisms is demonstrated by the common problem of attempting to lose weight. People often fight themselves over weight loss. It is, as recent brain research indicates, a futile and useless fight, for humans love food. The search for the edible and the delicious is constant in history, and continues now with new cuisines, new restaurants, new food crazes. In their book *Consuming Passions: The Anthropology of Eating,* Peter Farb and George Armelagos comment:

> Humans will swallow almost anything that does not swallow them first. The animals they relish range in size from termites to whales; the Chinese of Hunan Province eat shrimp that are still wriggling, while North Americans and Europeans eat live oysters; some Asians prefer food so putrefied that the stench carries for dozens of yards. At various times and places, strong preferences have been shown for the fetuses of rodents, the tongues of larks, the eyes of sheep, the spawn of eels, the stomach contents of whales, and the windpipes of pigs.

Many pleasures are associated with food: one image of daily love and togetherness is a family gathered around a holiday table; the cliché has it that the way to a man's heart is through his stomach, and certain foods are considered aphrodisiacs; Jewish mothers are well known for attempting to cure almost every ailment with chicken soup.

Our love of and preoccupation with food has had, until recently, great adaptive value. In times when food supply was not reliable, people who gorged themselves when food was plentiful had a better chance of survival when it was not. In addition, when most work required the expenditure of enormous amounts of energy, huge meals were needed as fuel by the manual laborer. Until recently, there was no central heating, and in temperate climates, unheated homes made it necessary to produce the heat from within, from the products of the stomach rather than the tree and the furnace.

The body *itself* operates like a furnace, and the brain operates like the body's thermostat, but if more fuel in the form of calories* is taken

*A calorie is a measure of heat production, defined as the amount of energy required to increase the temperature of one gram of water by one degree Celsius.

in than is needed for heat or energy, it will be stored as fat. Thus, weight is gained when there are more calories taken in than are expended. However, losing and gaining weight is not simply a matter of calories consumed (as many diet books would have it), because the brain regulates weight around a set point, like the temperature set point on a thermostat.

Consider this: you will eat about fifty tons of food in your lifetime, yet once your growth has been achieved your weight rarely varies even as much as one ton! From this point of view, ten or fifteen pounds isn't very much. The set point is the body weight that the several brain areas attempt to maintain. The hypothalamus, for example, can control eating and drinking and metabolic level to raise or lower caloric expenditure. The set point thus keeps weight around a predetermined level. So it is more difficult than we would predict both to gain and lose weight by merely counting calories.

That the brain closely regulates body weight helps us to understand why it is easier to lose weight at the beginning of a diet and harder at the end. At the beginning we have just begun to move away from our set point and weight loss is relatively easy. As we lose weight we move far below our set point and loss becomes more and more difficult.

Consider these common excuses for being "overweight":

1. "I've lost hundreds of pounds in my life" (the implication being that it is always gained back).
2. "It doesn't matter how much I eat—I'm just naturally fat."
3. "I can gain weight *just looking* at food."

Recent evidence suggests that these clichés are true. Here is a fact that may make some of you sad: some people *are* born to be fat. The part of the brain that regulates weight may simply be set higher in some people. This makes losing weight below the point that is set difficult if not impossible.

The problem for the "constitutionally obese" is that their body norms, as expressed in their set points, are higher than the current cultural norm. The "naturally obese," then, may face bleak alternatives: either constant hunger or being considered overweight. This factor is why people lose and gain weight constantly and why most diets don't work. The dieters are up against a powerful biological barrier.

However, there is one consolation for people who seem doomed to never match the skinny ideal of the models in jeans ads. For years it has been assumed that thin people are healthier (most actuarial tables of insurance companies reflect this; obese people pay higher rates for life insurance). However, though many studies show that rats who are almost starved live the longest, it has recently been found that people who live the longest are well *above* the published norms for weight. In all age and height groups, people who are slightly fat are the healthiest, and the ideal weight for health increases with age. This increase is almost exactly the amount gained on the average. Our brain's automatic regulation in this case may well be wiser than our current cultural ideal, so being overweight may not be such a losing battle after all.

One reason it's hard to lose weight on diets is that the regulatory mechanisms of the brain begin immediately to compensate for the fuel shortage by conserving bodily energy. Dieters may recall that they often feel weak and sluggish after a couple of weeks on a diet, and even though fewer calories are being burned, there is less and less improvement in loss of weight. At the beginning of a diet when weight may well be near the set point, weight loss occurs fairly quickly. But as the weight is lowered, and the body strives harder to maintain its set point, metabolism is reduced, and weight is lost less quickly or not at all. The body's built-in protection against starvation makes the set point much higher than the person may desire. Because the body operates on a lower energy requirement during a diet, when you go off a diet and eat normally, weight gain is likely. So weight is kept in a delicate balance, regulated "unconsciously" by innate mechanisms of the brain.

There are stories told by many survivors of disasters that bear on all this. This one is from the Nazi concentration camps in World War II. Among their many unsavory experiments, the Nazis determined to discover how fast people would starve given inadequate diets. So they provided inmates with as few as three hundred calories per day. In one camp most of the inmates died on this diet, but one small group had many survivors. When the camps were liberated, the leader of the group was asked why he thought his fellow inmates survived. What did they do? He said, "Each day with our meager meal, we all gathered around and talked to each other. We talked about the most wonderful meals we had ever had, and the wonderful meals we might have, again, in the future. We imagined eating roasts and potatoes, cakes and wine."

Perhaps the brain can receive information about weight from sources other than food. That, too, is why looking at food may cause weight gain. In one study those who merely saw a sizzling steak increased insulin production. This increased the entry of fat into their cells. They *would* gain weight.

There may be mental and brain involvement in health far beyond our scientific ideas of only a few years ago. One further, and final, example comes from a recent and massive study of health. Jana Mossey and Evelyn Shapiro of the University of Manitoba studied three thousand people age sixty-five and older. Each person rated his or her own health from "poor" to "excellent." At the same time each person was also rated on health according to his or her medical records. The startling finding is this: those in objectively poor health who rated their health good had a *higher* chance of survival than those in objectively good health who rated their own health as poor. Although there may be many interpretations of this, it seems clear that what we *believe* about ourselves can allow us to conquer pain, change weight (and even survive), and it may well change our susceptibility to illness. Obviously there is much more to learn about the amazing control the brain exercises in health. Such discoveries may forever change the face of medicine and our knowledge of ourselves.

In a 1972 article in the *Bulletin of the Los Angeles Neurological Society,* distinguished neurosurgeon Joseph E. Bogen formally proposed the creation of a "giant walk-through brain." He envisaged a brain large enough for visitors to wander through its spaces, systems, and structures. It was to be a dramatic and enjoyable way of gaining an understanding of the complex organization of the brain's interior parts. In 1978, Bogen's proposal was given widespread exposure in his article for *Human Nature* magazine, illustrated by David Macaulay. What follows is an expanded version of the 1978 article, with increased emphasis on how such a "museum" might be designed and built as well as further elaboration of what visitors could expect to find as they wander through the intricacies of a giant brain.

A Modest Proposal

or

The planning, construction, and use of a giant brain for the edification and entertainment of us all

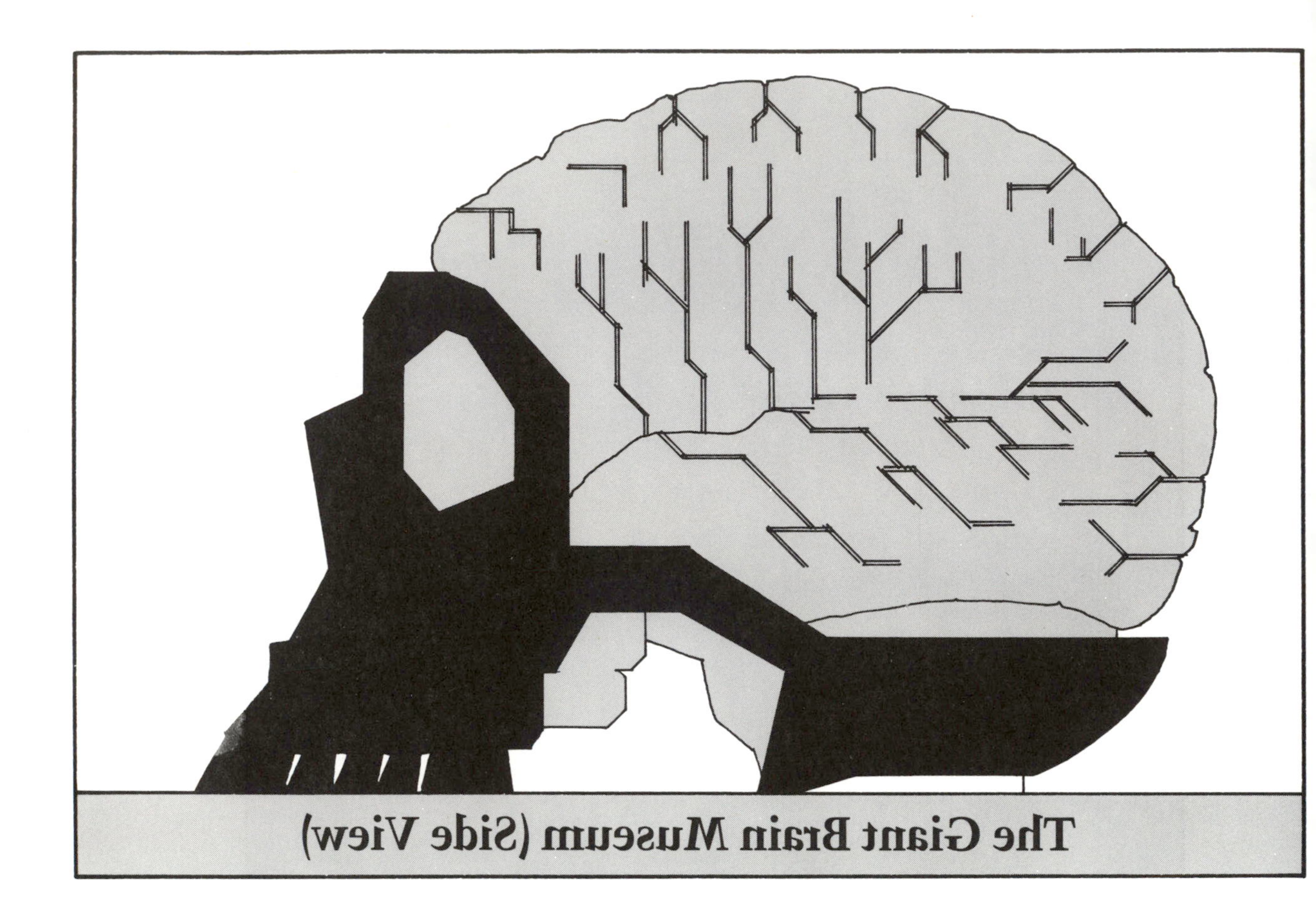

The Giant Brain Museum (Side View)

From the stem to the top of the cortex, the brain was to be almost five hundred feet tall and close to four hundred fifty feet long. It would be partially enclosed and supported by a large steel and concrete structure built in the form of a skull. Much of the wiring, piping, and ductwork serving the museum was to be secured to the exterior of the brain, as are the arteries and veins that supply blood to a living brain. Some of the larger interior arteries would also contain ramps and elevators.

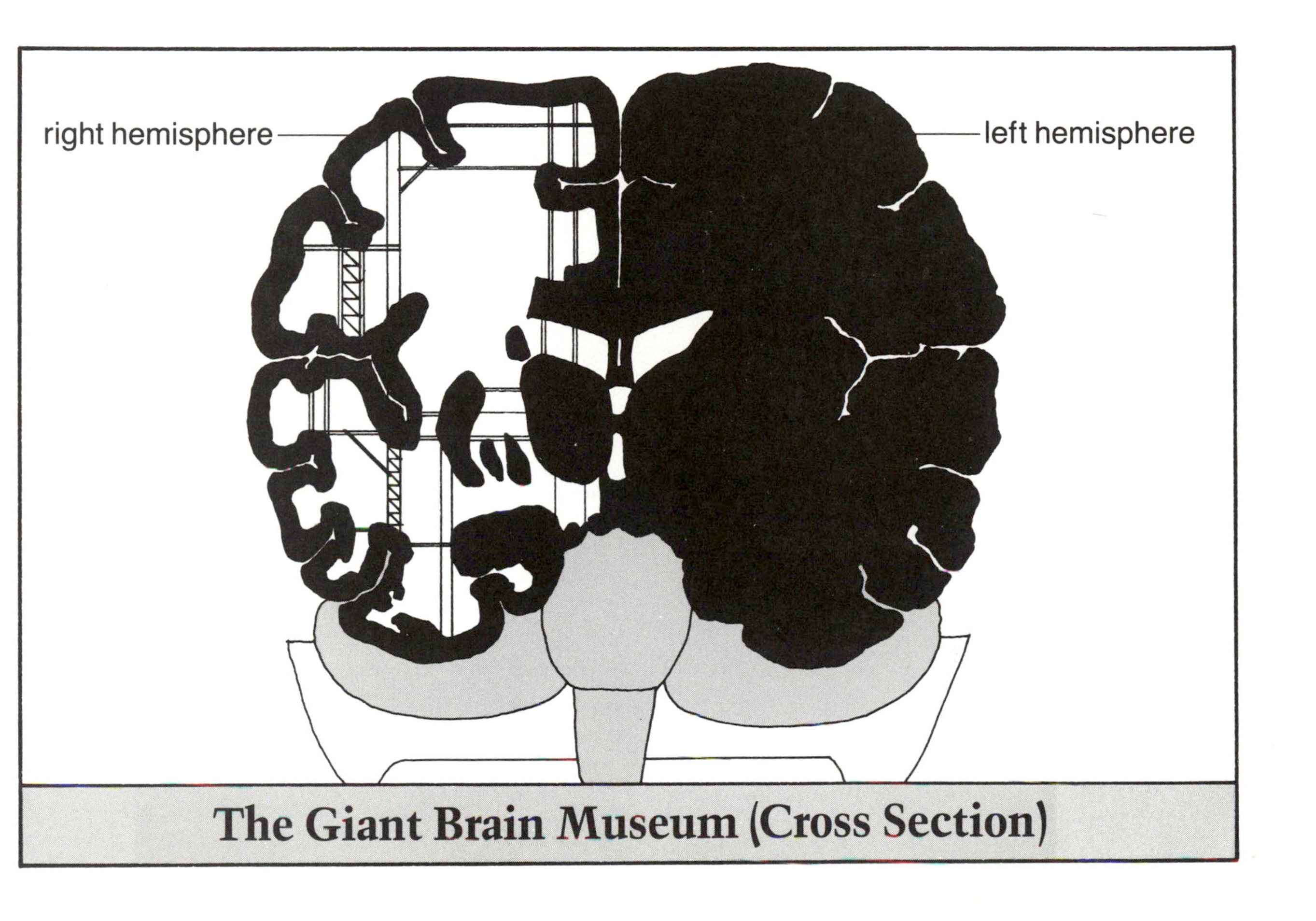

The Giant Brain Museum (Cross Section)

The left hemisphere would be almost solid, as in a living brain, and house various exhibitions, administrative offices, storage space, and—near the very top—a limited number of luxury condominiums. The income from the "cortical condos" would help to offset some of the museum's cost.

The right hemisphere was to be much more open, built as if most of the white matter had been removed. Thus visitors would be able to see gray matter structures, such as the thalamus and caudate nucleus, more clearly and to understand their relationship to one another better. In the open space above and around these structures, laser beams would delineate the various association and projection pathways that interconnect cortical areas of the same and opposite hemispheres.

Once the appropriate site had been selected from the many locations competing for the honor, constuction of the outer supports began.

As large areas of the partial skull neared completion, work began on the brainstem — the main entrance to the museum. Specially designed elevators would eventually carry visitors from the entrance area up through the pyramidal tract to the various exhibitions and tours in each hemisphere. (In the body, this tract runs all the way from the cortex to the lower spinal cord, and it is probably the route for all kinds of skilled movements.) Immediately behind the brainstem, the cerebellar auditorium was built. In it, introductory lectures and special presentations would be given.

While construction continued on the components of the limbic and visual systems of the right hemisphere, work was begun on the cortex of the left hemisphere.

For the next eighteen months, the slowly rising walls of both hemispheres were surrounded by an intricate web of scaffolding punctuated by a grove of tower cranes and derricks, their arms endlessly swinging back and forth.

Gradually the silvery form of the brain emerged from the entangled network of lines and motion. Large amounts of scaffolding were no longer required and only the huge safety nets, which hung just below the level of construction, remained attached to the vast convoluted surface. Within the year, both hemispheres were completely enclosed and most activity was now concentrated inside them.

By the fifth year, the remaining scaffolding, equipment, and work sheds were gone, and the army of workers was replaced by an even larger army of curious visitors.

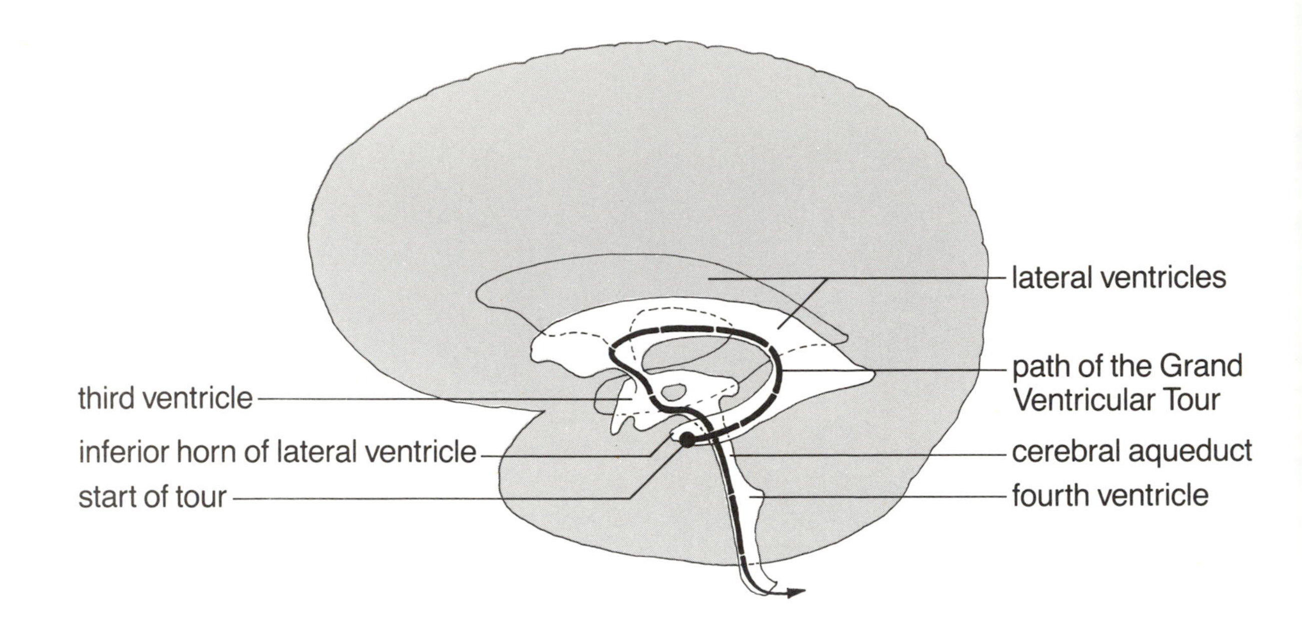

The Ventricles

Following an introductory presentation in the cerebellar auditorium, visitors set off for one or more of the spectacular tours. The two most popular offerings are the Grand Ventricular Tour and the Visual System Tour.

Ventricles are comparatively large spaces, three of which are located in the brain and one in the brainstem. The two largest ventricles—called the lateral ventricles—are symmetrically placed above the brainstem, one in each hemisphere. Since the lateral ventricle in the left, more solid, hemisphere is completely enclosed, it is there that the Grand Ventricular Tour begins. Visitors first gather in the inferior horn of the lateral venticle. After traveling along the ridges of the hippocampus, the group soon enters a ten-story-high space called the atrium. Here the horn curves gracefully upward and into the main body of the lateral ventricle. The ascent is breathtaking in every sense as tourists either climb the remainder of the hippocampus or wander up between the curious clusters of vessels that make up the choroid plexus. (In the living brain, the choroid plexuses produce cerebrospinal fluid, which normally fills all four ventricles as well as several other cavities.)

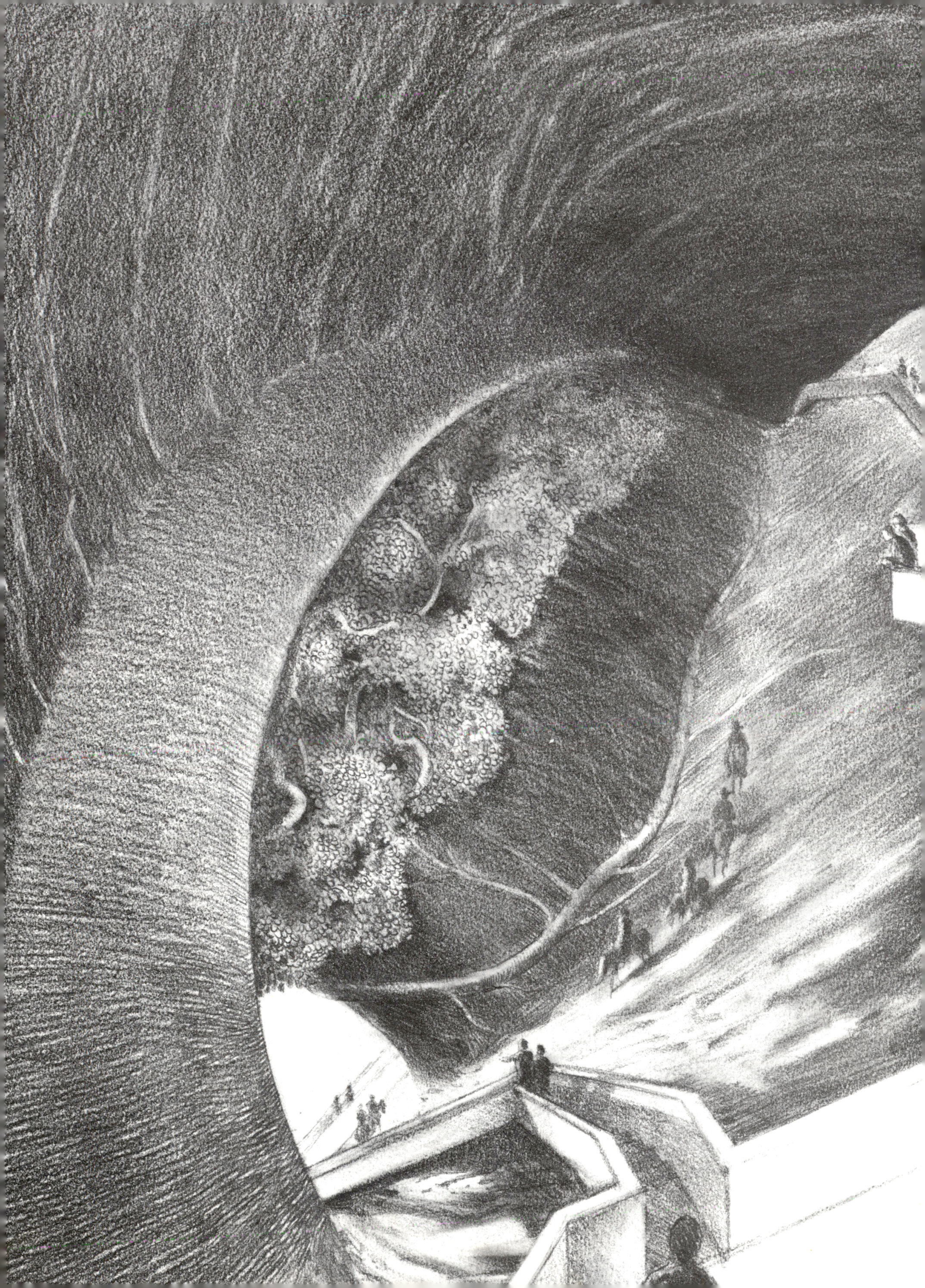

Eventually, all visitors enter the body of the lateral ventricle—a space almost the length of a football field. Upon reaching the front end of the ventricle, everyone visits the Caudate Café, both for refreshment and to enjoy the marvelous view. The café is built into the caudate nucleus, which serves as the outer wall of the lateral ventricle. From the verandah of the Caudate Café, visitors can look down toward the foramen of Monro, through which they will soon enter the third ventricle. Also passing through the foramen of Monro into the third ventricle are the choroid plexus and the internal cerebral vein, while above the opening stands the mighty arch of the fornix, which extends back along the entire length of the lateral ventricle. The ceiling of the lateral ventricle is formed by the transverse fibers of the corpus callosum, the principal connection between the two hemispheres.

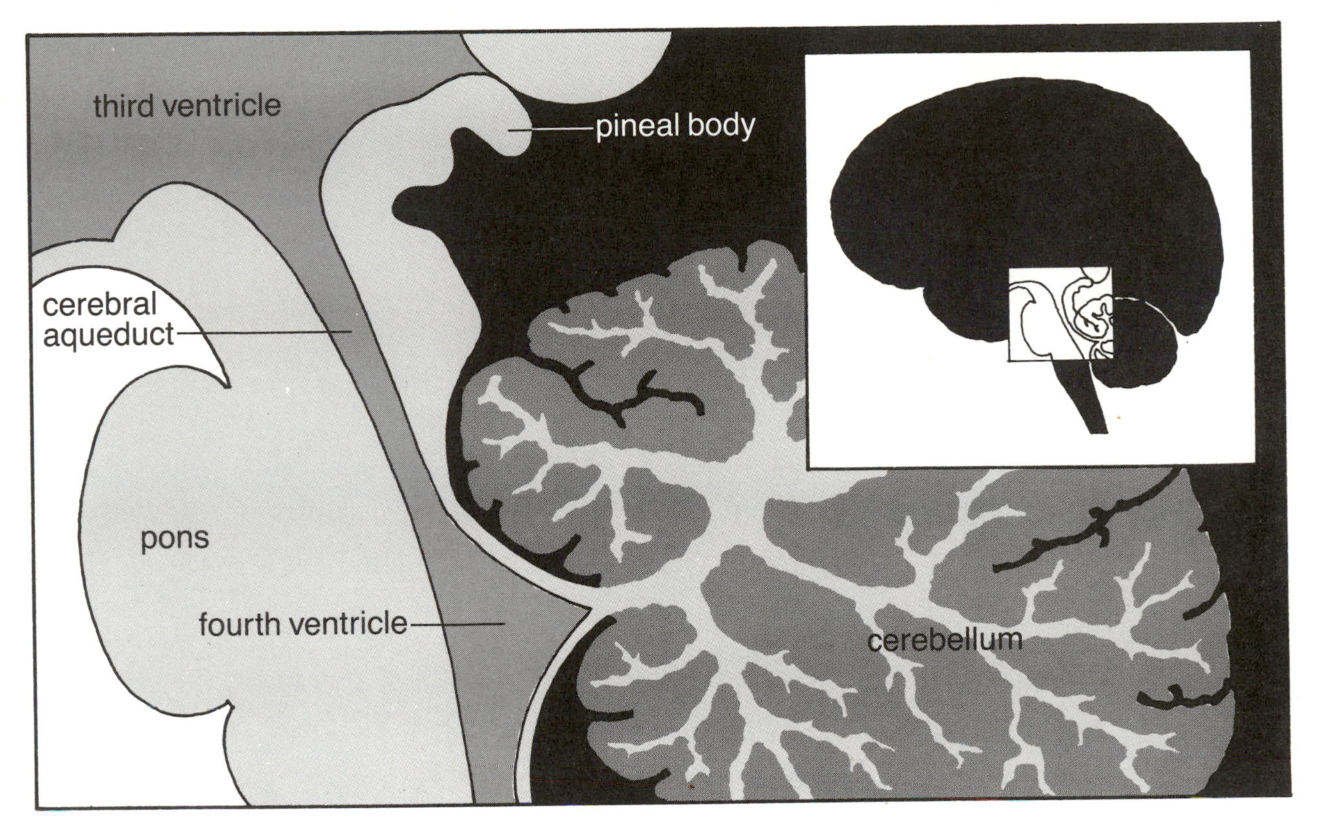

End of the Grand Ventricular Tour

The third ventricle is a space roughly in the middle of the brain that leads into the cerebral aqueduct. Arching over the entrance to the aqueduct is the posterior commissure—another band of fibers connecting the hemispheres. Above the posterior commissure is the pineal recess, into which some of the sightseers wander for a quick visit to the pineal gland—once thought to be the seat of the soul, it now serves as a rest room. Bridging the third ventricle is the massa intermedia—another route for interhemispheric communication, only this time between the two thalami. The Grand Ventricular Tour concludes with the descent through the narrow cerebral aqueduct (which normally carries cerebrospinal fluid to various cavities around the brain) and into the fourth ventricle.

Beginning of the Visual System Tour

The tour of the visual system begins in the eyes. Participants work their way back from the retinas, along the optic nerves to the optic chiasm. Depending on which of the two pathways they have chosen in each optic nerve, they either continue on to the thalamus on the same side of the brain as the eye from which they started or they cross over and travel to the thalamus of the opposite hemisphere. Both paths eventually lead to the visual cortex at the back of the brain.

Chiasmatic Cistern Seen from the Pituitary Stalk

Another, less ambitious, tour begins with an elevator ride up the carotid artery, stopping first at the inner ear and then at the space below the optic chiasm called the chiasmatic cistern. An aneurysm, an abnormal expansion of the carotid wall, is visible immediately upon leaving the elevator at the level of the optic cistern.

Just below the aneurysm, the posterior communicating artery stretches toward the back of the brain. Above the aneurysm, the anterior choroidal artery ramp ascends into the left temporal lobe. At the rear of the cistern, dropping down from behind the optic chiasm, is the pituitary stalk, which connects the hypothalamus with the pituitary gland. Above the cistern are the two glass-enclosed optic nerves leading to the eyes.

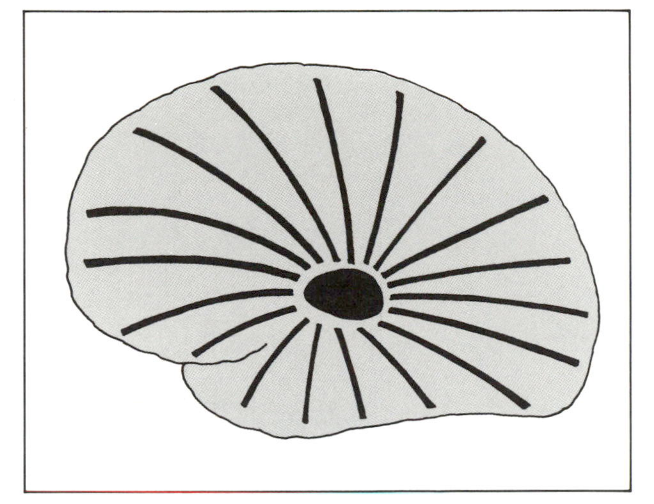

Projection Pathways from the Thalamus to Both Hemispheres

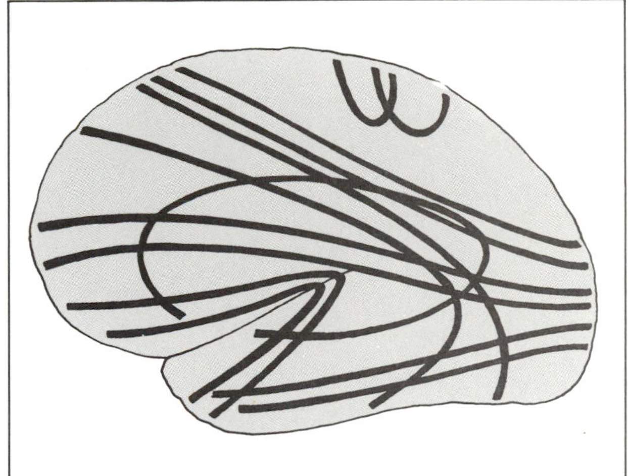

Association Pathways Within Each Hemisphere

At some point all visitors to the museum pass through the open space of the right hemisphere. Here laser beams are used to approximate either projection pathways (which extend between the thalamus and various parts of the cortex), association pathways (which interconnect various parts of the cortex within a single hemisphere), or the fibers of the corpus callosum (which connect left and right hemispheres). It is a spectacular display, particularly when viewed from one of the many catwalks that extend both around the interior surface of the cortex and across the space to offer closer inspection of the gray matter structures.

There are many other equally impressive tours and exhibitions available within the giant brain—far too many, in fact, to adequately describe here. Our visit, therefore, will conclude with a brief look at one of the museum's more unusual offerings.

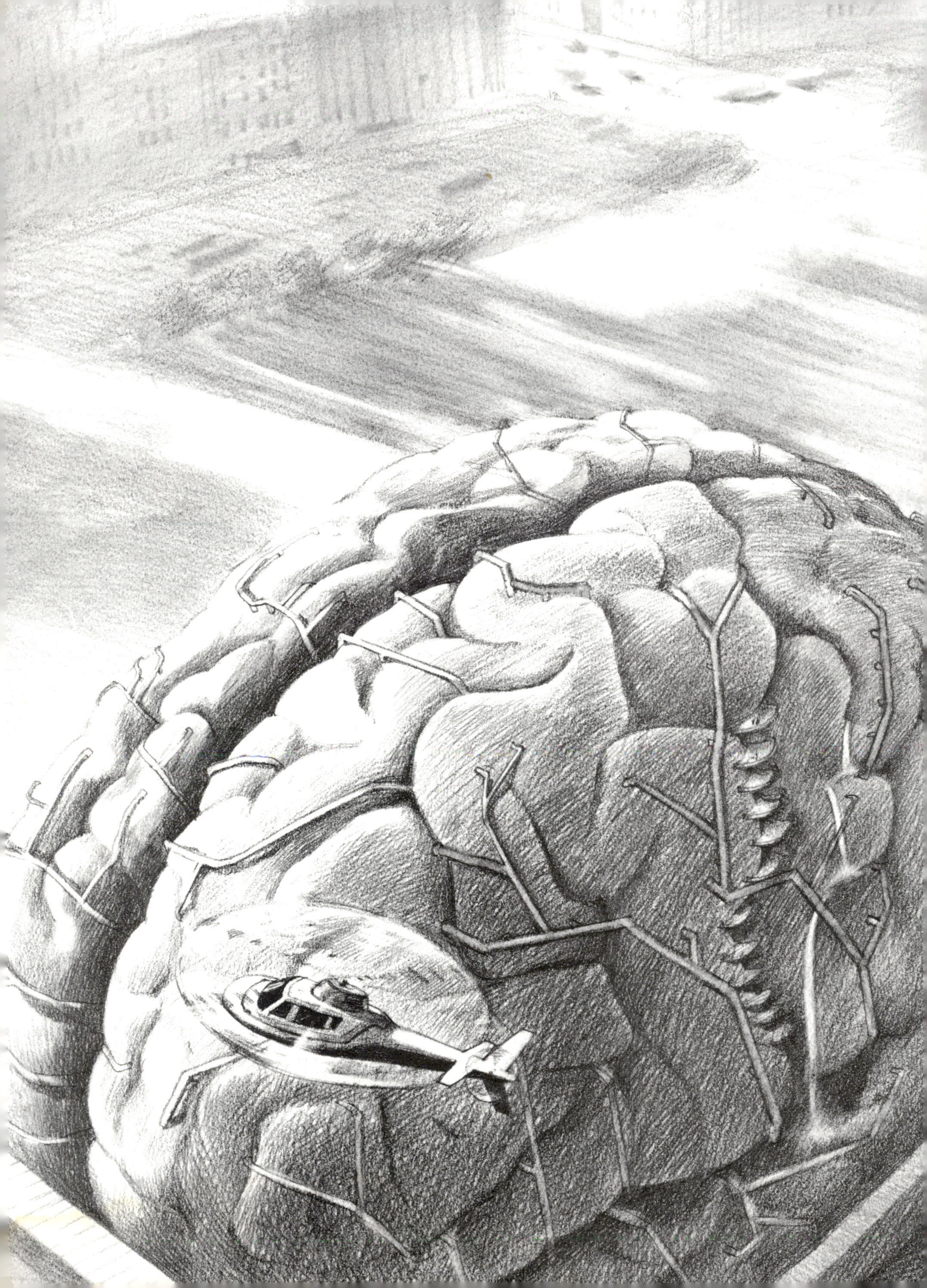

Cortical Condo Terrace

For some, the giant brain is considerably more than a dazzling event on a travel itinerary—it is home. A handful of affluent individuals who don't experience vertigo occupy the dozen luxurious condominiums tucked securely into the canyonlike fissure of Rolando. High atop the left parietal lobe, nestled between sensory and motor areas of the cortex, exists a tiny Shangri-la—a world of breathtaking views and luscious vegetation where the pressures of life are erased by the soothing roar of waterfalls splashing over the gyri (convolutions) and disappearing into the sulci (grooves), only to be magically and endlessly recycled.

But the idyllic environment created more than forty stories above their heads matters little to the majority of the giant brain's clientele. The magnificence of the museum's vistas and the incredible assortment of interconnecting structures fill everyone who sees them with wonder and awe, and perhaps more than just a little pride—for this is, after all, merely a simplified replica of each and every visitor.